T0214063

SpringerBriefs in Statistics

SpringerBriefs present concise summaries of cutting-edge research and practical applications across a wide spectrum of fields. Featuring compact volumes of 50 to 125 pages, the series covers a range of content from professional to academic. Typical topics might include:

- A timely report of state-of-the art analytical techniques
- A bridge between new research results, as published in journal articles, and a contextual literature review
- A snapshot of a hot or emerging topic
- An in-depth case study or clinical example
- A presentation of core concepts that students must understand in order to make independent contributions

SpringerBriefs in Statistics showcase emerging theory, empirical research, and practical application in Statistics from a global author community.

SpringerBriefs are characterized by fast, global electronic dissemination, standard publishing contracts, standardized manuscript preparation and formatting guidelines, and expedited production schedules.

More information about this series at http://www.springer.com/series/8921

Xiaomei Zhang • Guohong Cao

Event Attendance Prediction in Social Networks

 Springer

Xiaomei Zhang
University of South Carolina Beaufort
Bluffton, SC, USA

Guohong Cao
Pennsylvania State University
University Park, PA, USA

ISSN 2191-544X   ISSN 2191-5458 (electronic)
SpringerBriefs in Statistics
ISBN 978-3-030-89261-6   ISBN 978-3-030-89262-3 (eBook)
https://doi.org/10.1007/978-3-030-89262-3

This Springer imprint is published by the registered company Springer Nature Switzerland AG.
The registered company address is: Gewerbestrasse 11, 6330 Cham, Switzerland

Preface

Event attendance prediction has attracted considerable attention because of its wide range of applications. By predicting event attendance, events that better fit users' interests can be recommended, and personalized location-based or topic-based services related to the events can be provided to users. Moreover, it can help event organizers in estimating the event scale, identifying conflicts, and managing resources. This book first surveys the existing techniques on event attendance prediction and other related topics in event-based social networks. It then introduces a context-aware data mining approach to predict event attendance by learning how users are likely to attend future events. Specifically, three sets of context-aware attributes are identified by analyzing users' past activities, including semantic, temporal, and spatial attributes. This book illustrates how these attributes can be applied for event attendance prediction by incorporating them into supervised learning models and demonstrates their effectiveness through a real-world dataset collected from event-based social networks.

Bluffton, SC, USA   Xiaomei Zhang
University Park, PA, USA   Guohong Cao
May 2021

Contents

1   **Introduction** 1

2   **Related Work** 5
     2.1   Human Mobility Prediction 5
         2.1.1   Short-Term Mobility Prediction 6
         2.1.2   Long-Term Mobility Prediction 6
     2.2   Event-Based Social Networks (EBSN) 7
         2.2.1   Recommendation Systems 8
         2.2.2   Event Attendance Prediction 9
         2.2.3   Other Topics in Event-Based Social Networks 10

3   **Data Collection** 11
     3.1   Problem Definition 11
     3.2   Event-Based Social Networks and Data Collection 11

4   **Event Attendance Prediction: Attributes** 15
     4.1   Semantic Analysis 15
         4.1.1   Event Keywords 15
         4.1.2   Category-Based Similarity 16
         4.1.3   Categorizing Keywords 17
         4.1.4   Computing Event Similarity 18
         4.1.5   Discussions 19
     4.2   Semantic Attributes 20
         4.2.1   Topic Attendance 20
         4.2.2   Topic Loyalty 21
     4.3   Temporal Attributes 21
         4.3.1   Recent Attendance 22
         4.3.2   Weekly Preference 22
         4.3.3   Daily Preference 23
     4.4   Spatial Attributes 23
         4.4.1   Home Distance 24
         4.4.2   Location Preference 24

5   Event Attendance Prediction: Learning Methods 25
   5.1   An Overview of Data Mining 25
   5.2   Logistic Regression 26
   5.3   Decision Tree 28
   5.4   Naïve Bayes 29

6   Performance Evaluations 33
   6.1   Data Selection 33
   6.2   Experiment Setting 34
   6.3   The Effectiveness of Individual Attributes 35
   6.4   Supervised Learning 37
      6.4.1   Parameter Setup 38
      6.4.2   Comparisons of Learning Models 40
      6.4.3   The Effects of Semantic Information 42
      6.4.4   Effects of Training Set Length 44

7   Conclusions and Future Research Directions 47

References 49

Chapter 1
Introduction

Abstract This chapter introduces the topic of event attendance prediction and its potential applications. We introduce a data mining approach to predict event attendance, which has three research challenges, i.e., dataset collection, extraction of appropriate attributes, and identifying suitable learning methods. We first explain these challenges and then describe how to address them with a context-aware data mining approach. In this approach, three sets of context-aware attributes are used to characterize users' likelihood to attend future events, which are combined and fed into supervised learning methods to better predict event attendance.

Understanding and predicting human mobility can help solve problems in many different fields. For example, it can assist with many networking problems such as managing resources in cellular networks [1], developing location-based applications [2], and designing data dissemination protocols in mobile opportunistic networks [3–5]. In view of these applications, there have been a large amount of research on characterizing and predicting human mobility [6, 7]. One category of such research effort focuses on predicting users' short-term moving trajectories [8, 9], and another category of research focuses on predicting users' regular or periodic activities, i.e., predicting the periodic mobility pattern in future days or weeks [10, 11].

Although the aforementioned techniques can be applied to predict users' geographical mobility, they are not able to infer users' activities in many cases, such as during some social events, especially if the events are sporadic. When a sporadic event happens, such as an art festival in a plaza, it is hard to infer if a user will attend this event purely by predicting the user's mobility. Moreover, even though existing research can identify periodic visits to landmarks, they fail to capture periodic routines when there are location changes. For example, a class held every weekday morning may take place at different buildings. The weekend party of a group of students may be held at different places each week.

In this book, we address these problems by focusing on the prediction of event attendance, i.e., predicting users' attendance at a future event at specific time and location. The event can be a "English study group" where people with poor English skills gather together to practice English, or a "football seminar" where a famous

© The Author(s), under exclusive license to Springer Nature Switzerland AG 2021
X. Zhang, G. Cao, *Event Attendance Prediction in Social Networks*, SpringerBriefs
in Statistics, https://doi.org/10.1007/978-3-030-89262-3_1

football player gives a talk on campus. The *event attendance problem* can be formulated as a data mining problem which predicts whether a user will attend a future event by mining the user's past activities. Different from location prediction, where the same location can be revisited and future location can be inferred from past locations, events are not repeated. This is because each event is more likely associated with a new location and time, not simply a repetition of the past events. This problem is also referred to as the "cold-start" problem. To address this problem, we propose to identify "similar-topic" events in the past by studying the topic relevance between events, and then analyze users' activities on those past events. The basic idea is as follows: if a user is interested in events related to a topic, this user may also attend future events related to this topic. For example, if a student usually attends art-related events (e.g., art class on weekends), it is more likely he or she will go to the art festival in the coming weekend.

Based on this idea, we present a data mining approach which can learn how users are likely to attend future events, by addressing the following three research challenges.

1. How to collect an event based dataset for learning and predicting event attendance?
2. What attributes should be extracted to characterize how users are likely to attend future events?
3. What learning methods should be used to predict event attendance based on the extracted attributes?

To address the first challenge, we collect our own dataset due to the lack of any existing event based dataset. Thanks to the popularity of online social networks [12] such as Meetup (www.meetup.com), Plancast (www.plancast.com), and Eventbrite (www.eventbrite.com), where people in the neighborhood can create, organize and sign on social events online, we are able to collect event-based datasets. We have collected a dataset based on Meetup which contains 149,089 users and 132,739 events organized over a period of 2 years. The details of the data collection processes will be presented in Chap. 3.

To address the second challenge, we extract context-aware attributes to quantify users' likelihood of attending future events by analyzing their attendance to past events. The definition of context-aware attributes requires the analysis on the past events that have similar topics. To identify the topic relevance on events, a semantic analysis method is developed. Rather than simply counting the common keywords in the topic descriptions of events, the semantic analysis method calculates the "semantic similarity" between events. By analyzing the "semantic" meaning of the keywords in the descriptions, we are able to capture the topic relevance of the events more thoroughly. Based on the semantic analysis method, three sets of context-aware attributes are then extracted, including *semantic attributes*, *temporal attributes*, and *spatial attributes*. Semantic attributes focus on the event topics and characterize how often users attend similar-topic events in the past. Temporal attributes measure users' temporal preference when attending events, like weekly and daily preference on attending specific-topic events. Similarly, spatial attributes capture users' location preference when attending events.

To address the third challenge, we apply three supervised classifiers to learn how the extracted attributes affect users' decisions on event attendance, including *logistic regression*, *decision tree* and *naïve Bayes*. With the Meetup dataset, the performances of these classifiers are evaluated and compared. Evaluation results show that the classifiers built on multiple attributes significantly outperform those based on individual attributes, which indicates that event attendance can be affected by multiple factors. We also find that semantic attributes are more important than temporal attributes and spatial attributes when applied for event attendance prediction.

The rest of the book is organized as follows. Chapter 2 presents an review of the related work. Chapter 3 presents the data collection process. In Chap. 4, we introduce the semantic analysis methods and the context-aware attributes. Chapter 5 presents the learning methods and Chap. 6 presents the performance evaluation results. Chapter 7 concludes the book and points out future research directions.

Chapter 2
Related Work

Abstract This chapter describes some related research similar to that appearing in this book. Since much of the previous study focuses on human mobility prediction, the initial discussion centers around this topic. Existing works on short-term mobility prediction and long-term mobility prediction are reviewed. Then, we survey related work on event-based social networks, with focuses on recommendation systems and event attendance prediction.

2.1 Human Mobility Prediction

Instead of predicting users' attendance at regular or irregular social events, human mobility prediction focuses on predicting people's spatial mobility among locations. Mobility prediction can produce different types of outputs based on specific requirements. For example, we can predict the moving direction of a person within a limited geographical area, e.g., the driving direction in a city. We can also predict the next location of a person, for example, where the person will visit after work. In more complicated scenarios, there is a need for predicting the complete moving trajectory, e.g., FBI may be interested in finding out where the wanted person has gone.

Characterizing and predicting human mobility not only contribute to the understanding of users' spatial movement behaviors, but also help solving many networking problems such as handoff management [13], resource management [1], and location management [2, 14]. Therefore, it has attracted considerable attention in recent years.

Similar to event attendance prediction, the human movement history can be analyzed for predicting human mobility. Existing research [15–17] has shown that a person's future mobility can be predicted by analyzing his/her past movement. In this way, this individual's behavior can be fully characterized and exploited for predicting his or her future movement. However, there is some weakness with this approach. If a person does not have enough movement history recorded, the prediction will be inaccurate. To address this problem, researchers have proposed to analyze and exploit the common mobility characteristics from all users' movement history [18, 19]. Utilizing other users' movement history is especially helpful when there is not enough data. A better approach is to integrate these two perspectives, i.e., analyzing features of an individual user's movement, while at the same time considering the common features of all users' movement history. This integrated approach has been studied in [20].

Based on the mobility history, there are two kinds of prediction techniques: mobility prediction in a short term, e.g., predicting locations within the next few hours, and predicting regular activities in a long term, e.g., predicting the periodic mobility pattern within the next few days or weeks.

2.1.1 Short-Term Mobility Prediction

Short-term mobility prediction mainly focuses on characterizing users' historical spatial trajectories. Based on the mobility history, mobility models can be built, which are then used to infer future locations. The mobility models are usually built based on techniques like Markov chain, Hidden Markov model, and neural networks. These techniques can be used to model the transition probabilities among spatial locations, based on the users' movement history.

Regular Markov chain techniques have been exploited in predicting users' mobility [21, 22]. Markov chain can be used to formalize the transition probabilities from the current state (location) to another state (location). Different from the Markov chain techniques that calculate the transition probabilities between states, hidden Markov models assume the underling state transitions are hidden but can be inferred from the observed location trajectory, in which the location trajectory is the "outcome" of the hidden state transitions. Many techniques based on hidden Markov model have been proposed [15, 23, 24]. By adding a hidden layer of states, hidden Markov model based mobility prediction methods usually outperform Markov chain based approaches. Inspired by hidden Markov model, multi-layer neural network-based techniques have been recently proposed by researchers to infer user or vehicle mobility [8, 9, 25]. These methods are able to capture short term mobility. However, they usually fail to predict user mobility in a longer term (in days or weeks). This is because there will be a large amount of location transitions (states) in the longer term. Dealing with such large numbers of states will be much harder and then reducing the prediction accuracy.

2.1.2 Long-Term Mobility Prediction

In addition to the short-term mobility, there has been a large amount of research on the analysis of spatial-temporal patterns of user movement, i.e., the periodic mobility patterns, which is useful for characterizing the long-term human mobility. For example, a person regularly commutes among home, work place and gym in weekdays, forming a daily pattern. This person may also regularly visit the grocery store close to his or her home each Saturday afternoon, forming a weekly pattern.

Based on these observations, researchers [10, 11, 26, 27] have studied human periodic mobility pattern by analyzing the daily and weekly movement history. Sadilek et al. [28] further predicted human mobility in a longer term at the scale of months or years. Such research mainly utilizes data mining-based techniques to perform pattern mining on human behaviors. They are able to capture human mobility most of time, because human behaviors show high degree of repetition and periodicity. However, they fail to detect users' irregular movements such as at the irregular social events, which needs the help of event attendance prediction.

2.2 Event-Based Social Networks (EBSN)

Different from traditional online social media, the newly emerging online *Event-Based Social Networks (EBSN)* such as Meetup (www.meetup.com), Plancast (www.plancast.com), and Eventbrite (www.eventbrite.com), provide a new venue to organize and share offline social events. The *offline social events* are events hosted for people with shared interests, so that they are able to meet in person at specific time and location. Those EBSNs enjoy the benefits of both online social networks and offline social communities. First, users are able to connect to each other, organize events, and search for social groups and events in the online platform. Second, by attending the offline social events, they are able to build and maintain social relationship, and form offline social communities in real life.

Offline social events can be conveniently organized online, and users interact with each other face-to-face in the organized offline events. These networks also provide valuable information for researchers to study the online/offline human social behaviors. For example, Liu et al. [12] investigated the network properties of EBSN such as the degree distribution and community structures. Researchers in [29, 30] analyzed how offline activities at events affect the social networking behaviors of the attendees.

In EBSN, people can organize social groups and events for people with similar interests. Studying and knowing how to successfully organize groups and events are especially beneficial to the group organizers and event hosts. Many researchers studied the successful organization of groups and events in EBSN. Richen et al. [31] first discussed the possible reasons to start a social group and host social events in Meetup, and then analyzed the factors that lead to the success of social groups. Researchers [32, 33] also predicted the popularity and failure of social groups by studying and utilizing the group-related and event-related features. Feng et al. [34] studied how to search for an influential organizer for event, and Li et al. [35] further studied how to effectively and efficiently organize social events for a large group of people from a computational point of view. The goal is to maximize the "overall happiness" of the users by assigning users to various events being offered. Zhang et al. [36] further predicted the popularity of events by considering diverse influential factors, and discussed how to improve the event popularity. Other researchers [37, 38] discussed similar topics on how to promote events and maximize the influence of these events in social networks.

To maximize the success probability, organizers can actively invite users to groups and events. Actively inviting users not only helps expand the group size, but also helps enable users finding the emerging groups and events matching their interests. However, the challenge is how to identify who to invite from a large group of people. To address this challenge, researchers in [39] designed an algorithm to invite users considering their preferences based on their existing groups, while at the same time maximizing their influence to attract more attendees. Ai et al. [40] discussed an efficient way to invite users based on the historical data of their smart devices.

A subproblem of event organization is how to plan and arrange the upcoming events for a group of participants, which is called the problem of event-participant arrangement. Many researchers [41–46] have studied this problem and proposed solutions according to the events' minimum and maximum capacity requirement [43, 46], the users' interest [44], time availability [41], feedbacks [42], and social activeness [45].

In addition to the aforementioned work, other researchers try to improve the user experience in EBSN through group recommendation, event recommendation, and event attendance prediction, as detailed below.

2.2.1 Recommendation Systems

Similar to popular recommendation systems such as book recommendation and movie recommendation, recommendation in EBSN aims to recommend groups and events to users according to their interests. Whether an EBSN is successful or not usually depends on the recommendation system, because a good recommendation system helps expand the social network by connecting many potential users.

More specifically, group recommendation aims to recommend social groups to users according to their potential social interests. For example, Liu et al. [47] identified the group preference profile by considering the group members' personal preferences and personal impacts, so that it can be recommended to users with similar preference. Yuan et al. [48] used similar ideas to estimate the preference of a group by aggregating the preference of the group members, while giving different weights to the group members according to their influences. Zhang et al. [49] further considered the location features of the event-based groups and exploited the matrix factorization to model the relationship between users and groups. Other researchers [50] applied a collaborative filtering-based Bayesian model to capture multiple event semantics and group dynamics such as user interactions, user group membership, and user influence for personalized group recommendations.

Event recommendation is similar to group recommendation, but also involves more detailed characterization of the event, such as the time, location, and potential attendants. Different from traditional recommendation problems, event recommendation is more challenging since it has the unique cold-start nature [51]. This is because events are typically short-lived, with no historical attendance information available. To address this problem, researchers exploited history events that users attended to identify the potential interests of users, and the potential social influence between users. Based on this idea, existing event recommendation approaches mainly exploit event-based factors [52–56] such as time, location, and context-aware event description, and user-based factors [57–59] such as social influence of event hosts and group members to rank the events for personalized recommendations. To further study the social influence, Mariescu-Istodor et al. [60] pointed out that the social network connections can be potentially used for activity recommendations. Fränti et al. [61] studied whether the user connections (like the similarity on location history) on social networks can be applied for recommendation, but they concluded that the similarity on location history can only provide limited assistance for recommendation. Moreover, Mahajan et al. [62] considered real-time event recommendation by utilizing the IoT-edge-cloud data. Pham et al. [63] proposed a general graph-based model to solve three recommendation problems in one framework, including event recommendation, group recommendation, and recommending tags.

In addition to group and event recommendation, researchers [64] presented a deep-learning based venue recommendation system which provides context-driven venue recommendations for the event hosts to host their events. Lu et al. [65] studied the friend recommendation problem considering that an effective friend recommendation can help to promote user interaction and accelerate information diffusion for promoted events, and proposed a Bayesian latent factor approach for friend recommendation. Yin et al. [66] proposed a joint recommendation approach that not only recommends events but also recommends the partners at events, to promote event attendance.

2.2.2   Event Attendance Prediction

Another major research topic in EBSN is the analysis and prediction of users' event attendance behaviors. Researchers in [67–69] conducted analysis on users' behaviors. Specifically, Karanikolaou et al. [67] studied the factors that may influence users' behaviors in social networks and found that users typically only visit a few places frequently, and event types and topics usually play an important role on users' decisions. Zhao et al. [68] modeled users' preference when facing conflicted events. Trinh et al. [69] analyzed users' behaviors by characterizing their loyalty to the social groups and the hosted events, and how long they usually stay in a group.

Based on the analysis of users' behaviors, researchers proposed techniques to predict event attendance. Zhang et al. [70] applied data mining models to predict users' attendance to future events. The data mining models incorporate features that characterize the users' interests on event topics, and the temporal and spatial preference of the users. Du et al. [71] considered the social influence of event hosts based on the observation that users tend to attend events hosted by the same host. Xu et al. [72] further analyzed the mutual influence between users and studied how the dynamic social connections affect users' social gathering. Similarity, Wu et al. [73] studied who usually attend events together. Considering active users and inactive users may behave differently when attending events, Lu et al. [74] differentiated active users from inactive users and proposed a multi-features model that trains different feature parameters for active and inactive users respectively. Zhang et al. [75] studied how weather conditions impact event attendance and offline activities.

2.2.3   Other Topics in Event-Based Social Networks

In addition to the above topics that received considerable attention, some other topics are worth to mention. Researchers in [76] studied the possibility of privacy leakage in Meetup. The privacy issue is important in EBSN, since the users' on-site information is highly relevant to their real lives. This work analyzed what private information can be inferred from the site's publicly available data based on the crawled data in Meetup. They found that some specific social information such as the LGBT status, can be predicted with high accuracy. They also discussed the possible cause of privacy leakage and the possible damages.

Researchers also studied how local event meetups can be used as an offline forum to share knowledge in a specific field. For example, researchers in [77] discussed how software practitioners can use the informal local meetups to share software engineering knowledge. They conducted interviews and surveyed the leaders of technology oriented Meetup groups and found that participants in these groups use Meetup to stay abreast of new developments, build local networks, and share the cutting-edge knowledge with peers to improve their skills.

Chapter 3
Data Collection

Abstract Data is needed to solve any prediction or learning problems, and thus data collection is the first step for event attendance prediction. The data requirement usually depends on the definition of the specific problem. In this chapter, we first define the problem of event attendance prediction, and then describe how data is collected from an event-based social network.

3.1 Problem Definition

The problem of event-attendance prediction is defined as follows:

Definition Given a user and a future event, the problem of ***event attendance prediction*** is to predict whether the user will attend the event with the following information:

1. The information (description, time, location) of the future event.
2. The information (description, time, location) of the past events that the user has attended.
3. The home location of the user.

Based on the problem definition, the dataset collected should include all information (e.g., description, time, location) of the events hosted in a specific geographic area within a specific time period, as well as the basic information (e.g., home location) of the users in that geographical area. Fortunately, most event-based social networks provide repositories for events-based data, and our dataset is collected based on an event-based social network that has been popularly used today.

3.2 Event-Based Social Networks and Data Collection

During this modern age of social media, people are part of the social community. Online planning of community events can speed up information dissemination, and hence people use facebook.com, meetup.com and eventbrite.com to organize

© The Author(s), under exclusive license to Springer Nature Switzerland AG 2021   11
X. Zhang, G. Cao, *Event Attendance Prediction in Social Networks*, SpringerBriefs
in Statistics, https://doi.org/10.1007/978-3-030-89262-3_3

community events. Such online social networks enable people to create social groups, organize social events, and track the number of attendance. These networks are also referred to as event-based social networks [12, 70] because of their focus on event organization. Using meetup.com as an example, an upcoming event is posted in Meetup after it is planned by the hosting group. People reply "yes" if they would like to attend this upcoming event, and become the event attendants. From the website, we can also find information about the past events and people who attended those events.

Collecting data from online social networks seems bulky and hard to do. Fortunately, most social networks provide Application Program Interface (API) [78], through which data about events and their members can be collected and analyzed by researchers. API provides an easy to use interface, which allows users to pull information stored on the originators' database, and serves as the common interface in the technology and marketing industry. The API methods allow programs to connect easily and share information instantaneously.

The work presented in this book is based on the data collected through meetup.com. Meetup not only has many events with different topics, but also provides easy to use APIs for data collections. Meetup.com has its own set of API methods to gather information about social groups, events, attendance, and locations. Figure 3.1 shows the API instruction page given by meetup.com. Basically, the API provides a set of core methods and a common request format. They are combined to form a URL that returns the requested information, through HTTP. For example, Fig. 3.2 shows an example of the API call that lists social groups near the area with zip code 16801, and Fig. 3.3 shows an API call to retrieve all events hosted by a specific group.

In these two URLs:

1. `api.meetup.com` represents the API host.
2. `/find/groups` represents the method that is called. It is shown as a path, representing that the API is requesting the group information.
3. The parameters in this API, such as `zip`, `radius`, and `category`, are used to limit the search results by geographic location and area of interest.
4. `order` indicates how to order the results, in this case by the number of members.
5. `central-pennsylvania-observers-meetup` indicates the "url name" of the group from which the events are retrieved.
6. `page` indicates the maximum allowed entries in one page is 20.

Based on the above two methods, we can collect the event data. We first retrieve all groups in the local area and then collect past events hosted by each group. Similarly, data on users can also be collected through other API methods provided by Meetup.

Data storage using database (e.g., MySQL) allows for optimized retrieval and sorting. After requesting the information based on the URL, the raw data returned through API is in JavaScript Object Notation (JSON) format. JSON cannot be easily read or sorted by human eyes. Thus, we develop Java code to convert JSON data and import it into a local MySQL database. After being imported into MySQL database,

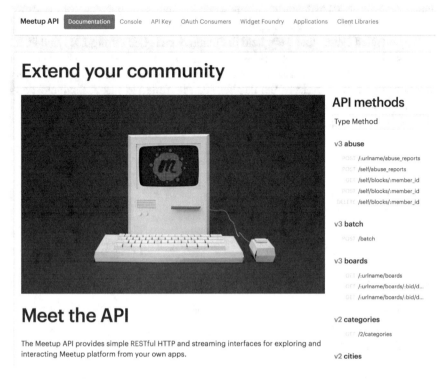

Fig. 3.1 The API instruction page provided by meetup.com, and the API is based on HTTP. The API methods provide interfaces for collecting data related to groups, events, members, categories, cities, venues, etc.

https://api.meetup.com[1]/find/groups[2]?zip=16801&radius=1&category=25[3]
&order=members[4]

Fig. 3.2 The API call to request for groups near zip code 16801

https://api.meetup.com/central-pennsylvania-observers-
meetup[5]/events?page=20[6]

Fig. 3.3 The API call to request for all events hosted by a specific group

data can be retrieved efficiently. In this work, a database is developed to store the information retrieved from meetup.com.

We collected events that were held in Pennsylvania (US) during 2011 and 2012. The resulting dataset contains 149,089 users having 132,739 events organized over 2 years. The collected information is stored in the MySQL database. The Entity Relationship (ER) diagram of the database can be found in Fig. 3.4, which shows the relationships of entity sets stored in the database.

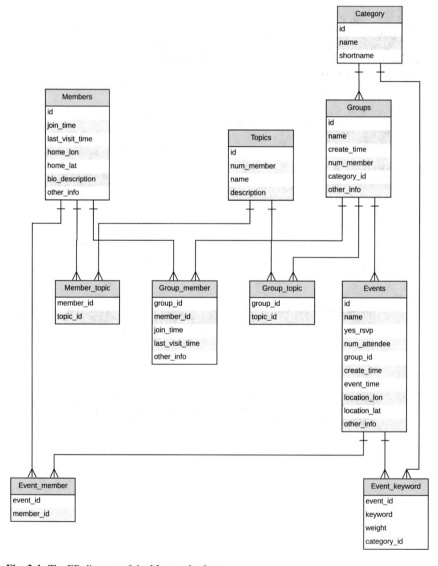

Fig. 3.4 The ER diagram of the Meetup database

Chapter 4
Event Attendance Prediction: Attributes

Abstract To characterize users' likelihood of attending a future event, attributes should first be extracted from the data, and then used in the learning methods to predict the event attendance. This chapter focuses on identifying the context-aware attributes. The definition of context-aware attributes requires analysis of past events with similar topics. Therefore, we first present a semantic analysis method to calculate the semantic similarity between events, and then explain the three sets of context-aware attributes in details, i.e., semantic attributes, temporal attributes, and spatial attributes.

4.1 Semantic Analysis

4.1.1 Event Keywords

Predicting users' attendance at events requires a semantic representation for each event so that the semantic similarity between events can be computed. To semantically represent an event, a keyword list is extracted from the description of the event: $\{k_1, k_2, ..., k_m\}$, where m is the number of keywords. Each keyword k_i is associated with a weight w_i $(0 \leq w_i \leq 1)$ which represents the importance of this keyword. In our database, the event keywords are extracted from the topics of the hosting group and the event description (the event title). A weight of 0.8 is assigned to the keywords extracted from the topics of the hosting group, and a weight of 1 is assigned to the keywords extracted from the event title. The keywords extracted from the event title have higher weight because they contain more unique information related to the event.

Consequently, an event can be semantically represented by a set of keyword pairs:

$$\{(k_1, w_1), (k_2, w_2), ..., (k_m, w_m)\}$$

Since the topic of an event can be represented by a set of keywords, we compute the similarity between two events as the similarity between two sets of keywords. A popular method to compute the similarity is to count the common keywords or compute the Jaccard similarity coefficient [79] between the two sets. These methods are implemented by matching the identical keywords in the two sets. However, they fail to identify the synonymies with similar semantic meanings. For example, the synonymies "speech" and "talk" are considered to be different, but in fact they have similar semantic meaning.

To address this problem, we consider the semantic meanings of the words, and calculate the word similarity based on the similarity of their semantic meanings. Some popular word similarity measures have been proposed based on WordNet [80, 81], which is a large lexical database of English not only providing the word definitions, but also recording various semantic relations between words. In this book, we adopt one of the most well-known word similarity measures—*Lin* [82] measure, which computes the similarity based on the word definitions. Other similarity measures are also tested via experiments, but only show minor difference from the *Lin* measure. Therefore, the *Lin* measure is used in this book.

4.1.2 Category-Based Similarity

After calculating the word similarity between keywords, an intuitive way to calculate event similarity is to sum the pairwise similarities of all keywords for the two events. However, summing all the pairwise similarities is still problematic, because the number of keywords in the events can impact the value of the similarity inappropriately. For example, an event might be given many keywords with similar meanings, like "food", "cuisine" and "cooking". Using all keywords will overestimate the event similarity. An alternative method is to only consider the maximum similarity on keywords. However, this method ignores the similarity of events in multiple categories. For example, two events on "women soccer" and "girls' sport" should consider the similarity on both categories of "women" and "sports", and only calculating the maximum similarity will underestimate the event similarity.

Considering these problems, the event similarity is calculated based on a category-based method. Specifically, the keywords of an event are first matched into categories, i.e., one keyword should belong to one category. The similarity at each category is then computed. The overall similarity is computed by summing the similarities in all categories. Then, as long as the categories represent different topics, the event similarity on distinct categories can be included, and the above problems can be addressed. In event-based social networks, the category information is usually pre-defined. For instance, Meetup provides the category information to categorize the topics of groups and events, and there are 33 categories as listed in Table 4.1, which will be used in this book.

Table 4.1 Categories of Meetup groups and events

Categories		
Arts/culture	Career/business	Cars/motorcycles
Dancing	Education/learning	Community/environment
Fitness	Food/drink	Fashion/beauty
Games	Government/politics	Lesbian/gay/bisexual/transgender
Hobbies/crafts	Health/wellbeing	Language/ethnics
Lifestyle	Movies/film	Literature/writing
Music	Spirituality	Outdoors/adventure
Paranormal	Parents/family	Pets/animals
Photography	Religion/beliefs	Fiction/fantasy
Singles	Socializing	Sports/recreation
Support	Technology	Women

4.1.3 Categorizing Keywords

To match a keyword to a category, the *word similarity* between the keyword and each category is computed, and this keyword belongs to the category that has the largest word similarity. Figure 4.1 shows an example of matching event keywords into categories. This event is about a swimming class for women, and it has two keywords: "swimming" and "women". Each keyword is matched to the category that has the largest word similarity. The keyword "swimming" has word similarity of 0.797, 0.619, and 0.611 with the categories "Sports", "Dance", and "Games", respectively, and is therefore categorized to "Sports", with the largest similarity. Similarly, the keyword "women" has the largest word similarity with category "Women", and therefore is categorized to the category "Women".

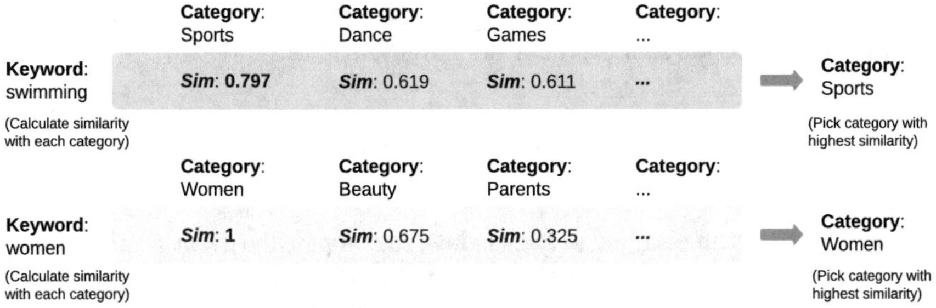

Fig. 4.1 Matching keywords "swimming" and "women" into categories

4.1.4 Computing Event Similarity

After matching keywords into categories, the set of keyword pairs is split into multiple subsets corresponding to all categories. The subset of keyword pairs in category i is denoted as $c_i = \{(k_{i,1}, w_{i,1}), (k_{i,2}, w_{i,2}), ...\}$, where $k_{i,u}$ is the u-th keyword and $w_{i,u}$ is the weight of $k_{i,u}$. To compute the similarity between two events e_1, e_2, we first calculate the event similarity in each category and then sum them. The method to calculate event similarity is similar to that proposed in [83], which calculates the semantic similarity on two sets of keywords by greedily matching the most similar keywords. As an improvement, our method further considers the similarity on distinct semantic categories, so that the similarity on different topics can be captured.

For the two events e_1 and e_2, $c_{i,1}$, $c_{i,2}$ represent the two subsets of keywords pairs in category i. Then, the event similarity in category i is the similarity between $c_{i,1}$ and $c_{i,2}$, denoted as $S_i(c_{i,1}, c_{i,2})$. It is calculated as the largest pairwise similarity between keywords in $c_{i,1}$ and $c_{i,2}$, where the similarity between two keywords is the word similarity between two keywords multiplied by their weights. The event similarity in category i is represented as follows:

$$
\begin{aligned}
S_i(c_{i,1}, c_{i,2})) \\
&= \max_{u,v}(KeywordPair((k_{i,u}^1, w_{i,u}^1), (k_{i,v}^2, w_{i,v}^2))) \\
&= \max_{u,v}(Word(k_{i,u}^1, k_{i,v}^2) * w_{i,u}^1 * w_{i,v}^2)
\end{aligned}
\tag{4.1}
$$

where $KeywordPair(*, *)$ represents the similarity between two keyword pairs, and $Word(*, *)$ represents the word similarity between two words. $k_{i,u}^1$ represents the u-th keyword in $c_{i,1}$, $k_{i,v}^2$ represents the v-th keyword in $c_{i,2}$, and $w_{i,u}^1$, $w_{i,v}^2$ are their weights.

Based on the similarity in each category, the *semantic similarity* between two events is defined as the summation of similarities in all categories:

$$
S(e_1, e_2) = \sum_{i=1}^{n} S_i(c_{i,1}, c_{i,2})
\tag{4.2}
$$

Figure 4.2 uses an example to explain how to compute the event similarity. One event has keyword pairs (swimming, 0.8) and (women, 1) and the other has (hiking, 0.8), (walking, 1) and (photos, 0.8). The similarity in the following categories are calculated.

- In "Sports" category, both events have keywords, and the event similarity in this category can be calculated using Eq. (4.1). For the keyword pairs (swimming, 0.8) and (hiking, 0.8), they have similarity $0.367 * 0.8 * 0.8 = 0.235$, where 0.367 is the word similarity between keywords "swimming" and "hiking", calculated

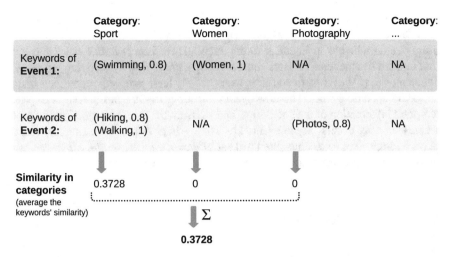

	Category: Sport	**Category**: Women	**Category**: Photography	**Category**: ...
Keywords of **Event 1:**	(Swimming, 0.8)	(Women, 1)	N/A	NA
Keywords of **Event 2:**	(Hiking, 0.8) (Walking, 1)	N/A	(Photos, 0.8)	NA
Similarity in categories (average the keywords' similarity)	0.3728	0	0	

$$\Sigma$$

0.3728

Fig. 4.2 Computing event similarity by summing similarities in all categories

based on the *Lin* measure. The two 0.8s are the weights of the keywords. Similarly, the keyword pairs (swimming, 0.8) and (walking, 1) have similarity $0.466 * 0.8 * 1 = 0.3728$. The keyword pairs (swimming, 0.8) and (walking, 1) have the largest similarity 0.3728, and thus the event similarity in category "Sports" is 0.3728.

- In the "Women" category, only Event 1 has keyword in this category, and therefore the similarity is 0.
- In the "Photography" category, only Event 2 has keyword in this category, and therefore the similarity is 0.

Adding the similarity in all categories, the semantic similarity between these two events is 0.3728.

4.1.5 Discussions

By calculating the semantic similarity on events based on the method presented in this section, we are able to find events with similar topics. This is especially important for event attendance prediction. By analyzing users' activities in the past events that have similar topics, we are able to estimate users' likelihood of attending future events. For example, if a user attended similar events frequently in the past, it is more likely that the user will attend such events in the future. We can also further consider the event time, e.g., if the user frequently attended similar events on Saturdays in the past, it is with very high chance that the user will still attend such events on Saturdays. In the rest of this chapter, we present the details on how to use event similarity to define the context-aware attributes.

In general, three sets of context-aware attributes are extracted to predict event attendance. The topic of the event is usually the major factor to consider when users decide whether to attend a future event. Therefore, *semantic attributes* are first defined to characterize users' interest in specific topics, by studying the users' attendance to past events. In addition, users usually have temporal preference when attending specific events. Therefore, *temporal attributes* are introduced to exploit temporal information of user activities, and measure the users' temporal preference when attending events. Since events are usually held offline, event location also plays an important role. Thus, the third set of attributes, i.e., *spatial attributes*, are defined to measure users' location preference when attending events.

4.2 Semantic Attributes

With semantic attributes, we aim to capture the user's interests in the events of a specific topic by studying his or her attendance to past events. For example, if a student frequently attended soccer games in the past, it is most likely that he will appear at a soccer game on campus or attend a seminar from a famous soccer player visiting campus. To quantify user u's interest in the topic of a future event e, two semantic attributes *topic attendance* and *topic loyalty* are introduced.

4.2.1 Topic Attendance

The *topic attendance* attribute measures the number of similar-topic events (or simplified as similar events) the user has attended in the past. The method introduced in Sect. 4.1.4 can be used here to measure the similarity between two events. Then, we use a parameter **similarity threshold** θ, to decide whether two events are similar, i.e., two events are considered to be similar events if their semantic similarity is larger than θ. The parameter θ can be flexibly set and its settings will be discussed in detail in Chap. 6.

Given event e, topic attendance is the number of similar events that u has attended during a time period T in the past. Here, **the past time period** T is a parameter determining how the event attendance history should be inspected. Its settings will also be discussed in Chap. 6.

Then, the first semantic attribute, *topic attendance*, can be calculated as follows:

$$x_{s_1} = |\{e_i \in E_u : t_{e_i} \in (t_e - T, t_e) \wedge S(e, e_i) > \theta\}| \tag{4.3}$$

where E_u indicates the set of past events that u has attended, and t_e is the time of event e.

4.2.2 Topic Loyalty

Topic loyalty is a semantic attribute used to measure users' "loyalty" to a specific topic. A user is considered to be loyal to a topic if he or she has attended all or most of the events of this topic. More specifically, topic loyalty is computed as the percentage of the events that the user has attended among all the events with similar topics. This attribute is effective when the events of a specific topic are only held infrequently and the user attends most of them. Although topic attendance is not able to capture users' interest in this topic due to the low frequency, *topic loyalty* is able to effectively identify users' interest in such events.

Specifically, for a future event e, we find all similar events held in a past time period T and calculate the percentage of them that u has attended. Formally, the total number of similar events in a time period T is as follows:

$$|\{e_i \in E : t_{e_i} \in (t_e - T, t_e) \wedge S(e, e_i) > \theta\}|$$

where E is the set of all past events. In formula (4.3), we computed the number of events that user u has attended in the past, i.e., $|\{e_i \in E_u : t_{e_i} \in (t_e - T, t_e) \wedge S(e, e_i) > \theta\}|$. Then, the *topic loyalty* for user u and event e is computed as:

$$x_{S2} = \frac{|\{e_i \in E_u : t_{e_i} \in (t_e - T, t_e) \wedge S(e, e_i) > \theta\}|}{|\{e_i \in E : t_{e_i} \in (t_e - T, t_e) \wedge S(e, e_i) > \theta\}|} \tag{4.4}$$

4.3 Temporal Attributes

Users usually have temporal preference when attending events. For example, a user may prefer to go to gym in the afternoon after work, and a user may prefer to watch football games on Saturday. To characterize the temporal preference of a user u on a future event e, three temporal attributes are extracted: *recent attendance*, *weekly preference*, and *daily preference*.

To define a temporal attribute, the basic idea is to calculate how a future event e is temporally related to similar-topic events that u has attended in the past. We use $R(e, e_i)$ to indicate the *temporal relation* between e and a past event e_i. Then, a temporal attribute is generally calculated by averaging the temporal relations between e and all the past events with similar topics:

$$x_t = \frac{\sum_{e_i \in E_u^T, S(e, e_i) > \theta} R(e, e_i) S(e, e_i)}{\sum_{e_i \in E_u^T, S(e, e_i) > \theta} S(e, e_i)} \tag{4.5}$$

Here, $e_i \in E_u^T$ indicates each of the past events that user u has attended in time period T. $R(e, e_i)$ is set according to various types of temporal relations as discussed in the following subsections, e.g., $R(e, e_i)$ becomes $R^r(e, e_i)$ when *recent attendance* is used, and it becomes $R^d(e, e_i)$ when *daily preference* is used. When averaging the temporal relations, the event similarity $S(e, e_i)$ between e and e_i, is used as the weight of the past event e_i, so that the past event with higher similarity to e has more contribution in computing the temporal attribute.

4.3.1 Recent Attendance

Human behaviors can be better characterized by the most recent activities. For example, if a user has frequently attended physical training recently, it is more likely that the user will attend another physical training event in the near future. Based on this observation, we propose an attribute called *recent attendance*, which can be defined by extending formula (4.5). The temporal relation $R(e, e_i)$ used in the formula should be able to quantify how close e to e_i, and thus should be related to the time interval between them. We use the reciprocal of the time interval as the temporal relation between the events:

$$R^r(e, e_i) = \frac{1}{t_e - t_{e_i}}.$$

Then, a more recent event has higher value of temporal relation. Based on formula (4.5), *recent attendance* is formulated as follows:

$$x_{t_1} = \frac{\sum_{e_i \in E_u^T, S(e,e_i) > \theta} R^r(e, e_i) S(e, e_i)}{\sum_{e_i \in E_u^T, S(e,e_i) > \theta} S(e, e_i)} \tag{4.6}$$

4.3.2 Weekly Preference

The second temporal attribute aims to quantify user's weekly preference. This is because people tend to have similar activities on the same day of the week. For example, a user prefers to attend football events on Saturday and go to farmers' market on Sunday. Similarly, *weekly preference* is defined by extending formula (4.5). The temporal relation $R(e, e_i)$ used in the formula should indicate if events e and e_i happen on the same day of the week. In particular, the temporal relation $R^w(e, e_i) = 1$ if e and e_i happen on the same day of the week, and $R^w(e, e_i) = 0$ if they happen on different days of week.

Based on formula (4.5), *weekly preference* attribute is calculated as follows:

$$x_{t_2} = \frac{\sum_{e_i \in E_u^T, S(e,e_i) > \theta} R^w(e, e_i) S(e, e_i)}{\sum_{e_i \in E_u^T, S(e,e_i) > \theta} S(e, e_i)} \tag{4.7}$$

4.3.3   Daily Preference

The third temporal attribute characterizes another human routine, i.e., they tend to have similar activity at similar time of a day. For example, a person normally attends the after-work exercise class around 5 pm every day. Similarly, *daily preference* is defined by extending the formula (4.5), and the temporal relation $R(e, e_i)$ used in the formula should be able to indicate if events e and e_i happen at the same time of the day. One thing to note here is that users' schedule may not be exactly the same everyday, i.e., it may be a little bit early or late when attending the same activity. Considering this, a Gaussian function is applied to measure how two time points (in hours) in a day are similar:

$$s(t_1, t_2) = e^{-\frac{(t_1 - t_2)^2}{2}}$$

The function returns a value lower than or equal to 1. The closer the two time points (t_1 and t_2) are, the larger the value is. Based on this formula, the temporal relation for *daily preference* is calculated as

$$R^d(e, e_i) = s(t_e, t_{e_i}) = e^{-\frac{(t_e - t_{e_i})^2}{2}}$$

The *daily preference* attribute is then calculated as:

$$x_{t_3} = \frac{\sum_{e_i \in E_u^T, S(e,e_i) > \theta} R^d(e, e_i) S(e, e_i)}{\sum_{e_i \in E_u^T, S(e,e_i) > \theta} S(e, e_i)} \tag{4.8}$$

In addition to the above temporal attributes, additional temporal attributes can be considered for some specific types of events. For example, the season factor [60] can be considered for outdoor events (e.g., bicycling, skiing, etc.), i.e., users usually go bicycling during spring and fall, and only go skiing during winter.

4.4   Spatial Attributes

Users may attend events based on the location of the event, and hence spatial attributes are extracted to capture users' spatial preferences. Next, we introduce two spatial attributes: *home distance* and *location preference*.

4.4.1   Home Distance

Home distance simply measures the distance between the home location of user u and the location of event e. Let L_u^h represent the home location of user u, and Let L_e represent the location of event e. Then, $dist(L_u^h, L_e)$ is the distance between the two locations. The first spatial (location-based) attribute, *home distance*, is

$$x_{l_1} = dist(L_u^h, L_e) \tag{4.9}$$

For offline events, the location information is in the form of latitude and longitude. The distance on locations can be calculated based on their latitudes and longitudes [84].

4.4.2   Location Preference

This attribute characterizes the users' location preference. For example, a user may only attend football games held in a specific stadium, and may not attend if the game is held at other locations. We use a similar formula as in (4.5) to characterize how event e is spatially related to past events. Here, the spatial relation is the distance between the locations of e and e_i, i.e., dist(L_e, L_{e_i}). The *location preference* can be represented as follows:

$$x_{l_2} = \frac{\sum_{e_i \in E_u^T, S(e,e_i) > \theta} dist(L_e, L_{e_i}) S(e, e_i)}{\sum_{e_i \in E_u^T, S(e,e_i) > \theta} S(e, e_i)} \tag{4.10}$$

In addition to aforementioned spatial attributes, additional attributes can be considered for events that involve moving activity, e.g., the outdoor events. For example, the move type [60, 85], i.e., the category of moving activity (e.g., walking, running, bicycling, or driving), can be detected from users' GPS records and considered as spatial attributes. However, using the move type requires the recording of the users' location trajectories, which are usually not available in the event-based social networks (like Meetup).

Chapter 5
Event Attendance Prediction: Learning Methods

Abstract After all the attributes are extracted, a learning method is utilized to learn how the extracted attributes affect users' decisions on event attendance. This process is also referred to as supervised binary classification, considering that 'attend or not' is a binary classification. There are many supervised classifiers in the literature, and we adopt three classifiers, including logistic regression, decision tree and naïve Bayes. In this chapter, we first give an overview of the data mining process, and then present the details of these supervised classifiers.

5.1 An Overview of Data Mining

The general data mining process is shown in Fig. 5.1. First, the dataset (i.e., input) is generated from the database that contains the collected data (as shown in Fig. 3.4), with each data example representing a user-event pair $\{u, e\}$. Specifically, for each user-event pair $\{u, e\}$, we use $\mathbf{x} = \{x_{s_1}, x_{s_2}, x_{t_1}, x_{t_2}, x_{t_3}, x_{l_1}, x_{l_2}\}$ to represent the extracted attributes (as defined in the last chapter) and use $y \in \{0, 1\}$ to represent whether u eventually attends e or not. The attributes $\mathbf{x}$ and the class label y are incorporated into one data example in the dataset:

$$[\mathbf{x}, y]$$

A dataset is created by generating data examples from an adequate number of user-event pairs. As we can see from Fig. 5.1, the dataset is then split into training data and testing data. The training data is used to train the supervised classifier, and the testing data is used to evaluate the trained classifier. Based on this process, we can train multiple classifiers, evaluate their performances, and determine which classifier has the best result.

There are many supervised classifiers in the literature [86], and we adopt three classifiers, including logistic regression [87], decision tree (C4.5) [88] and naïve Bayes [89].

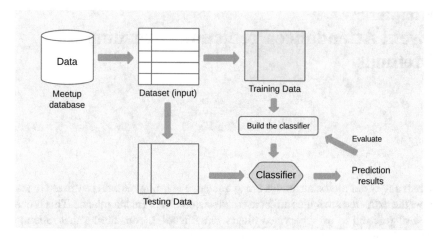

Fig. 5.1 The data mining process

5.2 Logistic Regression

When applied for classification, the logistic regression model is used to model the probability of a certain class. It is most commonly used for binary (Bernoulli) classification, like pass/fail, win/lose, alive/dead, healthy/sick, or attend/not attend an event. It can also be extended to multiple-class cases, such as determining what type of contact lenses should be recommended to patients, etc. In this case, each type of contact lens is assigned a probability between 0 and 1, with a sum of one.

To understand how logistic regression works, we consider a classification problem with n attributes $x_1, x_2, ..., x_n$, and a binary variable Y ($Y \in \{0, 1\}$) to represent the two classes. Let p denote the probability of $Y = 1$, $p = P(Y = 1)$. Logistic regression assumes that there is a linear relationship between the attributes $x_1, x_2, ..., x_n$ and the log-odd of value p. Let l_p denote the log-odd of p, i.e., $l_p = \ln \frac{p}{1-p}$, where ln is the log function with the base to be the natural number e. The linear relationship can be written as:

$$\ln \frac{p}{1-p} = \beta_0 + \beta_1 x_1 + \beta_2 x_2 +, ..., +\beta_n x_n$$

$$= \beta_0 + \sum_{i=1}^{n} \beta_i x_i \qquad (5.1)$$

where β_i represents the parameters, i.e., the coefficients of the attributes $x_1, x_2, ..., x_n$. The coefficients are learned by fitting the model to the training data, in which we know the values of attributes and the outcomes (i.e., the classes) of the data examples. The maximum likelihood estimation method [90] is used to learn the coefficients.

Once the parameters $\beta_1, \beta_2, ..., \beta_n$ are learned from the training data, the logistic regression model can be applied to the new data example in the test data to estimate the class Y. Based on Eq. (5.1), the probability p of $Y = 1$ is calculated as:

$$p = \frac{e^{\beta_0 + \sum_{i=0}^{n} \beta_i x_i}}{e^{\beta_0 + \sum_{i=0}^{n} \beta_i x_i} + 1}$$

$$= \frac{1}{1 + e^{-(\beta_0 + \sum_{i=0}^{n} \beta_i x_i)}} \tag{5.2}$$

Equation (5.2) is exactly in the form of a sigmoid function (also called logistic function) [91], $\text{sig}(t) = \frac{1}{1+e^{-t}}$. Therefore, we can simply write p as the sigmoid function of $\beta_0 + \sum_{i=0}^{n} \beta_i x_i$:

$$p = \text{sig}(\beta_0 + \sum_{i=0}^{n} \beta_i x_i) \tag{5.3}$$

Figure 5.2 shows the sigmoid function. The x-axis represents the value of $\beta_0 + \sum_{i=0}^{n} \beta_i x_i$ which ranges from negative infinity to positive infinity. The y-axis represents the probability p that the class variable Y is 1, and it ranges from 0 to 1. The probability that Y is 0 is simply $1 - p$. The logistic regression classifier classifies the data example to the class with higher probability.

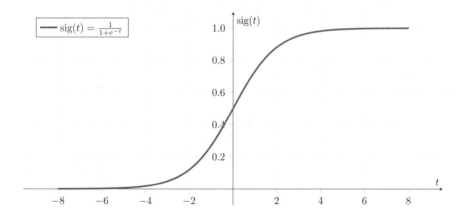

Fig. 5.2 The sigmoid function

5.3 Decision Tree

Decision tree is one of the most commonly used models for classification. Figure 5.3 shows how decision tree is used to determine what things to do in a particular day.

As we can see, a decision tree is a tree-like structure in which each internal node represents a "test" on an attribute (e.g., "work to do" is yes or no), each branch represents the outcome of the test, and each leaf node represents a class label, i.e., the final decision on the classes. The paths from root to leaf represent classification rules. We can also see from this figure that the decision tree provides good visualization of the classification rules, which is one of the reasons why decision tree is widely used.

Once a decision tree is built, it is easy to use by simply following the decision nodes from the root to the leaf. However, the most challenging process is to build up the tree. One widely used algorithm is C4.5 developed by Ross Quinlan [88] (Top ten Algorithms in Data Mining [92]). C4.5 is an extension of Quinlan's earlier ID3 algorithm [93].

We first explain how ID3 works and then extend it to C4.5. For a decision tree algorithm, the goal is to pick a good attribute to put on each node (from root to leaf) of the tree, so that by splitting on that attribute, the data examples can be better divided. For example, for a problem to determine the "commute time" to work (long, medium or short), we first need to decide which attribute should be put on the root of the tree. Assume there are two attributes "accident" and "weather" giving the following splits as shown in Fig. 5.4. In this case, "accident" is a better attribute to choose, because the subset in each branch is mainly in one class. However, the subsets after splitting on "weather" still have multiple classes in each. Therefore,

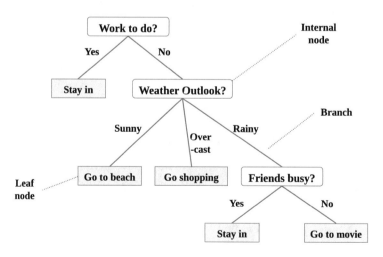

Fig. 5.3 A decision tree example to decide what things to do in a day

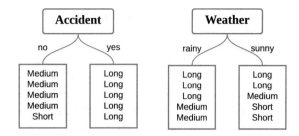

Fig. 5.4 The subsets after splitting on two attributes. The attribute of accident is preferred since it can better divide the data examples

"weather" is not able to divide the data well and is not considered to be a good attribute. The attribute to put on the root node is "accident".

Based on this idea, ID3 selects the best attribute using the concept of information entropy, a quantity to measure the information contained in a dataset. Generally speaking, entropy is a measure of the "diversity" of different classes in a dataset, and the goal is to make sure that after splitting on an attribute, the overall entropy (i.e., diversity) is reduced as much as possible at the branches. The reduction on entropy is also called information gain. Let E denote the entropy of the original dataset. The entropy after splitting on the i-th attribute is E_i, i.e., a weighted sum of entropy at all branches. The information gain of attribute i is:

$$IG_i = E - E_i. \tag{5.4}$$

Then, the attribute with the highest information gain is chosen to put on the node to make a decision. After making a decision on that attribute, the dataset is split into multiple subsets or branches. The ID3 algorithm recurses on each of the resulted branches and chooses attributes to build more nodes. The algorithm stops adding nodes to a branch until all the data examples on that branch belong to the same class, or none of the remaining attributes provides any information gain.

Based on the ID3 algorithm, the C4.5 algorithm adds several enhancements. One important improvement is that C4.5 can handle continuous attributes in addition to the discrete attributes which are the only attributes allowed in ID3. To handle the continuous attribute, C4.5 is able to perform a binary split based on a threshold. The dataset is split into two subsets, one for the data with attribute value above the threshold and one for the data with attribute value less than or equal to the threshold. Another important improvement is that C4.5 enables tree pruning. That is, C4.5 can go back through the tree once it has been created, and it removes branches that do not help and replaces them with leaf nodes.

5.4 Naïve Bayes

Naïve Bayes classifiers are a family of simple probabilistic classifiers based on applying Bayes' theorem. Similar to logistic regression classifier, naïve Bayes calculates the probabilities of the classes based on the attributes. However, they have a strong (naïve) assumption that the value of a particular attribute is independent of the value of any other attribute, given the class variable. For example, a fruit may be considered to be an apple if it is red, round, and about 10 cm in diameter. A naïve Bayes classifier considers that each of these attributes contribute independently to the probability that this fruit is an apple, regardless of any possible correlations between the color, roundness, and diameter attributes.

In general, naïve Bayes calculates the conditional probability of the classes given the observed attributes. Given an observed data example $\mathbf{x}$ with n attributes, $\mathbf{x} = \{x_1, x_2, ..., x_n\}$, naïve Bayes calculates the probability of this example to be in each class C_j:

$$p(C_j|\mathbf{x}) = p(C_j|x_1, x_2, ..., x_n) \tag{5.5}$$

and classifies $\mathbf{x}$ to the class C_j that has the highest probability $p(C_j|\mathbf{x})$.

Using Bayes' theorem, this conditional probability can be calculated as

$$p(C_j|\mathbf{x}) = \frac{p(\mathbf{x}|C_j)p(C_j)}{p(\mathbf{x})} \tag{5.6}$$

In this equation, the denominator $p(\mathbf{x})$ is usually hard to calculate, but it does not relate to the class C_j and can be considered as a constant z. Thus, in practice, only the numerator $p(\mathbf{x}|C_j)p(C_j)$ needs to be calculated. The data example $\mathbf{x}$ is simply classified to the class C_j that has the highest $p(\mathbf{x}|C_j)p(C_j)$.

In the numerator $p(\mathbf{x}|C_j)p(C_j)$, the probability of class C_j, $p(C_j)$, is calculated by simply dividing the number of data examples in class C_j by the total number of examples. To calculate $p(\mathbf{x}|C_j)$, the "naïve" conditional independence assumption needs to be used, i.e., assuming that all attributes $\{x_1, x_2, ..., x_n\}$ in $\mathbf{x}$ are mutually independent, conditional on the class C_j. Based on this assumption, we have

$$\begin{aligned} p(\mathbf{x}|C_j) &= p(x_1, x_2, ..., x_n|C_j) \\ &= p(x_1|C_j)p(x_2|C_j), ..., p(x_n|C_j) \\ &= \prod_{i=1}^{n} p(x_i|C_j) \end{aligned} \tag{5.7}$$

where $p(x_i|C_j)$ is the percentage of data examples with attribute value x_i from all data examples in class C_j.

Then, we have

$$p(\mathbf{x}|C_j)p(C_j) = \prod_{i=1}^{n} p(x_i|C_j)p(C_j), \tag{5.8}$$

and $\mathbf{x}$ should be classified to the class C_j that has the highest $\prod_{i=1}^{n} p(x_i|C_j)p(C_j)$.

The biggest disadvantage of the naïve Bayes classifier is the independent assumption on the attributes. However, despite its naïve design and apparently oversimplified assumptions, naïve Bayes has worked quite well in many complex real-world problems. One advantage of naïve Bayes is that it can be trained very efficiently. The naïve Bayes classifier can be built by a single scan of the database. Another advantage of naïve Bayes is that it only requires a small number of training data to build the classifier.

Chapter 6
Performance Evaluations

Abstract Based on the collected dataset, we train the classifiers and evaluate the effectiveness of the extracted attributes. We also evaluate the performance of the proposed solutions and evaluate how different parameters affect the performances. In this chapter, we first discuss the data selection, the experiment setting, and then present the evaluation results on the effectiveness of individual attributes and the performance of the three classifiers.

6.1 Data Selection

To evaluate the effectiveness of the extracted attributes and train the three supervised classifiers, the collected dataset is used. As discussed in the last chapter, each data example in the dataset includes the attributes and the class label (attend or not) of the user-event pair $\{u, e\}$, i.e., $[\mathbf{x}, y]$, where $\mathbf{x}$ represents the attributes, and $y \in \{0, 1\}$ represents the class label.

Data examples are selected from all events held during 2011–2012 in Pennsylvania. The users are those who attended at least one of those events. Figure 6.1 plots the distribution of the number of events each user attended during these 2 years. As can be seen from the figure, the majority of users have only attended one or two events, and only a small percentage of users (called *active users*) attend at least one event every month. Since inactive users do not attend events frequently, there is no need to include them for training the classifiers. Thus, only the active users are considered when selecting data examples. A data example is positive if $y = 1$, and it is negative if $y = 0$. To train the binary classifier unbiasedly, there should be equal numbers of positive examples and negative examples. In our dataset, there are 60,221 positive examples including all the active users and their attended events. To have the same number of negative examples, we randomly choose 60,221 negative examples from the active users and the events they have not attended.

© The Author(s), under exclusive license to Springer Nature Switzerland AG 2021

X. Zhang, G. Cao, *Event Attendance Prediction in Social Networks*, SpringerBriefs in Statistics, https://doi.org/10.1007/978-3-030-89262-3_6

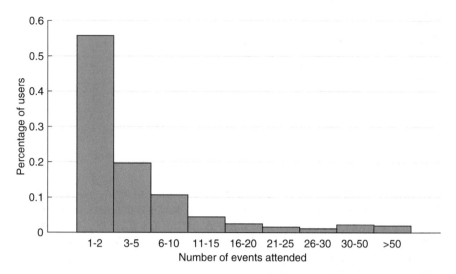

Fig. 6.1 The number of events attended by users within 2 years

6.2   Experiment Setting

The supervised classifiers used in our experiments include logistic regression, decision tree (C4.5) and naïve Bayes. One advantage of using these classifiers is that we are able to inspect the inner structures of these classifiers, so that we can learn the role of each attribute in prediction. The first half of the dataset is used for training and second half is used for evaluation.

The performance of various approaches is evaluated with the *Receiver-Operating-Characteristics (ROC)* [94] curve, which is commonly used to illustrate the performance of a binary classifier system. The ROC curve plots the fraction of true positives out of the total actual positives (true positive rate) vs. the fraction of false positives out of the total actual negatives (false positive rate), at various threshold settings. It is a monotonic non-decreasing function of true positive rate over the false positive rate. For example, for a *random* classifier that randomly classifies data to positive or negative, it results in a curve $y = x$ in the diagonal, because the true positive rate keeps the same as false positive rate. A classifier that is better than random will have a curve that is above diagonal, because the true positive rate is higher than the false positive rate. More generally, a classifier performs better if it is closer to the upper-left corner. Based on this, the *area under the ROC curve (AUC)* is an important metric to evaluate the performance of a binary classifier [95]. It is claimed in [96] that AUC is a statistically consistent and discriminating metric. Besides AUC, we also use a prediction *accuracy* metric, which is the percentage of correctly predicted event attendance examples.

6.3 The Effectiveness of Individual Attributes

We evaluate the prediction power of each individual attribute using decision stump, which is a one-level decision tree [97]. Based on each individual attribute, a predicting score needs to be first assigned for each user-event pair, and a higher score indicates a higher attending probability. For the semantic attribute or temporal attribute, a higher attribute value usually implies a higher attending probability, so the score is simply set to be the attribute value. For the spatial attribute, a smaller attribute value usually implies a higher attending probability, so the score is set to be the negative of the attribute value. By setting a decision threshold, the instance with score higher than the threshold is predicted as positive (will attend), otherwise it is predicted as negative (not attend). As the decision threshold varies, we obtain different true positives and false positives, which are used to generate the ROC curve. The ROC curves for the decision stumps built by individual attributes are shown in Figs. 6.2, 6.3 and 6.4. The reference line (diagonal) in the figures represents the ROC curve generated by the random classifier. To be more general, we evaluate the effects of attributes computed under different parameters: *similarity threshold θ* and *past time period T*. These two parameters were introduced in Chap. 4 to define the attributes. θ is the threshold to determine whether two events are considered to be "similar events". T determines how long the event attendance history should be inspected. Since the impact of T on the ROC curve is hard to observe, we only present the ROC curves under different θ, with $T = 90$ days.

Figure 6.2 shows the ROC curves for decision stumps built by semantic attributes: *topic attendance* and *topic loyalty*, under different settings of θ. We compare four approaches, *topic attendance* ($\theta = 1$ and $\theta = 3$) and *topic loyalty* ($\theta = 1$ and $\theta = 3$). As can be seen from Fig. 6.2, all four ROC curves are higher than the reference line, which means both semantic attributes are effective in predicting event attendance. We can also see that *topic loyalty* outperforms *topic attendance*.

Fig. 6.2 ROC curves for the decision stumps built by semantic attributes: *topic attendance* and *topic loyalty*. A classifier is better if its ROC curve is closer to the upper-left corner. The reference line in the diagonal represents the ROC curve for the random classifier. ©2015 IEEE. Reprinted, with permission, from X. Zhang et al. [70]

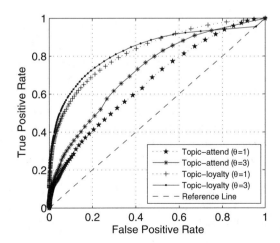

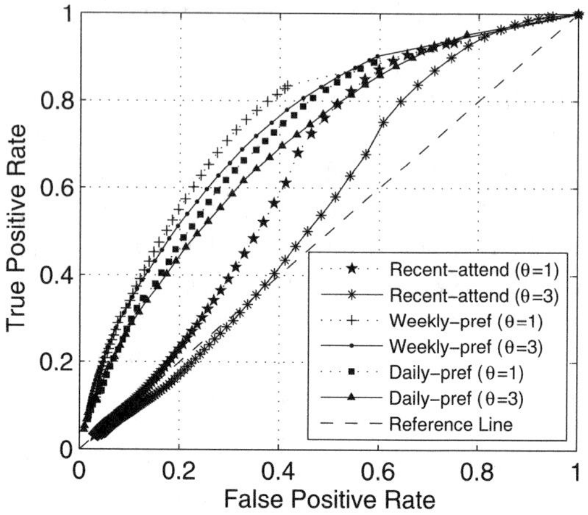

Fig. 6.3 ROC curves for the decision stumps built by temporal attributes: *recent attendance*, *weekly preference* and *daily preference*. ©2015 IEEE. Reprinted, with permission, from X. Zhang et al.[70]

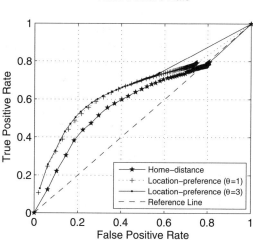

Fig. 6.4 ROC curves for the decision stumps built by spatial attributes, *home distance* and *location preference*. ©2015 IEEE. Reprinted, with permission, from X. Zhang et al. [70]

If a user is "loyal" to a topic, i.e., the user has attended all or most events of this topic, he is very likely to continue attending future events of this topic, and hence topic loyalty is a better attribute for predicting event attendance. In general, for both attributes, the curves of $\theta = 3$ are higher than those of $\theta = 1$ due to the following reason. With a threshold $\theta = 3$, only more similar events are considered in the calculation of semantic attributes. This implies that, rather than considering many unrelated events, only considering more similar events can better help infer the event attendance in the future.

Figure 6.3 shows the ROC curves for the decision stumps built by three temporal attributes: *recent attendance*, *daily preference* and *weekly preference*, under both $\theta = 1$ and $\theta = 3$. As can be seen, all the six ROC curves are higher than the reference line, demonstrating the overall effectiveness of the temporal attributes. We also observe that *weekly preference* outperforms the other two temporal attributes. This is because users' event attendance follows weekly pattern. By checking users' weekly attendance pattern, we are able to infer the future event attendance at each day of the week. Moreover, for all three temporal attributes, the curves of $\theta = 1$ are higher than those of $\theta = 3$. This result is different than those of semantic attributes. Setting a lower similarity threshold ($\theta = 1$) means more events are considered in the calculation of temporal attributes. If more events are considered, users' temporal patterns can be better characterized, which results in better temporal attributes. Therefore, to generate more effective temporal attributes, it is better to include more events by setting a lower similarity threshold θ.

Figure 6.4 shows the ROC curves for the decision stumps built by two spatial attributes: *home distance* and *location preference*. The similarity threshold θ only impacts *location preference*, and we have two ROC curves for *location preference* ($\theta = 1$ and $\theta = 3$), and one ROC curve for *home distance*. Similar to semantic and temporal attributes, the ROC curves for the two spatial attributes are higher than the reference line, indicating the effectiveness of the two spatial attributes. We can also find that the curves for *location preference* are higher than the curve of *home distance*. This is because users put preferences on event locations, i.e., if a user usually attended events at one location, he is likely to attend more events in this location or nearby. However, the distance to home is less important. For *location preference*, the two curves with $\theta = 1$ and $\theta = 3$ are very close, implying that the similarity threshold θ may not have large influence on spatial attributes. From the above analyses, we find that θ has difference impacts on different attributes. That is, $\theta = 3$ works better for semantic attributes, $\theta = 1$ works better for temporal attributes, and θ has negligible influence on spatial attributes. Since θ affects the overall performance, we will further study its impacts in Sect. 6.4.1

Finally, we compare the predicting power of all individual attributes by computing the AUCs corresponding to the ROC curves in Figs. 6.2, 6.3 and 6.4. As shown in Fig. 6.5, the results are consistent with what we have observed from the ROC curves, where *topic loyalty* outperforms other semantic attributes, *weekly preference* outperforms other temporal attributes, and *location preference* outperforms the other spatial attribute. Amongst all these attributes, *topic loyalty* has the best predicting power, with AUC higher than 0.7 for both $\theta = 1$ and $\theta = 3$. These results show that semantic attributes are the most important attributes for predicting event attendance.

6.4 Supervised Learning

In this subsection, we evaluate the performance of the supervised learning methods. We first set up the two parameters T and θ, and then evaluate and compare the performance of the three supervised classifiers.

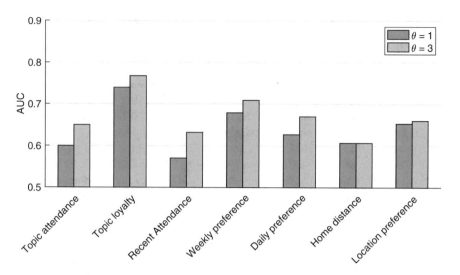

Fig. 6.5 AUC for individual attributes using decision stumps

6.4.1 Parameter Setup

Two parameters are used to define the attributes in Chap. 4, including the *past time period T* and the *similarity threshold θ*. The *past time period T* determines how long the event attendance history should be inspected, and the *similarity threshold* θ determines what events are considered to be "similar events". In Sect. 6.3, we evaluated the impacts of θ on the effectiveness of individual attributes. In this subsection, we setup these two parameters θ and T based on the evaluation results, considering different classifiers.

To evaluate how the parameters influence the performance, we train the super-vised classifiers using the attributes calculated with different T and θ, and then test their performances. The performances of the classifiers are shown in Figs. 6.6, 6.7 and 6.8 for the three classifiers *logistic regression*, *decision tree*, and *naïve Bayes*, respectively. In the evaluations, both AUC and prediction accuracy are used as performance metrics.

For the *logistic regression* classifier, as shown in Fig. 6.6, both AUC and accuracy reach maximum when $\theta = 2.5$, no matter what T value is used. This indicates that $\theta = 2.5$ is the best choice for the *logistic regression* classifier. By observing the different curves generated with different T, we find that $T = 30$ days has the worst performance for both AUC and accuracy, and $T = 60$ and $T = 90$ have much better performance. This is because $T = 30$ does not have enough history data to infer future event attendance. The difference between $T = 60$ and $T = 90$ is not significant. They have similar accuracy as shown in Fig. 6.6b, but $T = 90$ has slightly higher AUC than $T = 60$ as shown in Fig. 6.6a. We also increased T to be longer than 90 days (not shown in the figure), and could not find any noticeable

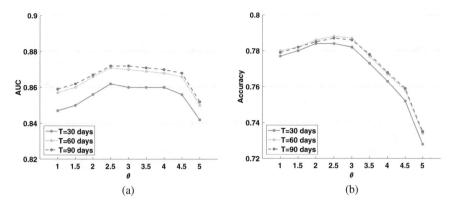

Fig. 6.6 The impact of parameters T and θ on performance with logistic regression. (**a**) AUC. (**b**) Accuracy

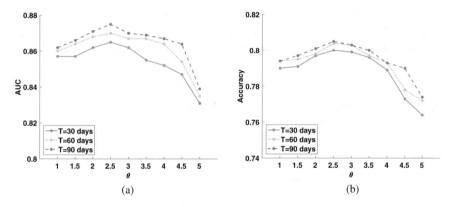

Fig. 6.7 The impact of parameters T and θ on performance with decision tree. (**a**) AUC. (**b**) Accuracy

change for AUC. These results suggest that 90 days of history data is enough to infer users' event attendance.

For the *decision tree* classifier, as shown in Fig. 6.7, both AUC and prediction accuracy reach maximum when $\theta = 2.5$, regardless of T. Therefore, θ should be set to 2.5 for *decision tree*. Under different values of T, we can see that $T = 90$ has better AUC and accuracy, and hence T is set to 90 days for the *decision tree* classifier.

For the *naïve Bayes* classifier, as shown in Fig. 6.8, both AUC and accuracy reach maximum when $\theta = 2.5$, regardless of T. Thus, θ is set to 2.5 for *naïve Bayes*. Under different values of T, *naïve Bayes* has similar accuracy, as shown in Fig. 6.8b. As shown in Fig. 6.8a, $T = 60$ and $T = 90$ have similar AUCs, which are much higher than that of $T = 30$. To be consistent with the other two classifiers, T is also set to 90 days for the *naïve Bayes* classifier.

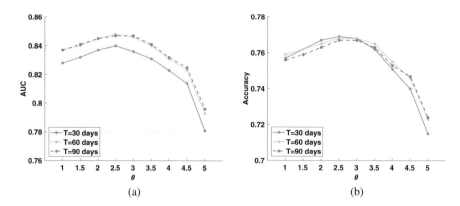

Fig. 6.8 The impact of parameters T and θ on performance with naïve Bayes. (**a**) AUC. (**b**) Accuracy

Based on the above analyses, θ is set to 2.5 and T is set to 90 days for all three classifiers in the following experiments.

6.4.2 Comparisons of Learning Models

In this subsection, we compare the performance of different classifiers, with various attributes. The evaluation results are shown in Table 6.1 using the AUC metric and shown in Table 6.2 using prediction accuracy. As can be seen, the classifiers built with all attributes can increase AUC by 0.02–0.25, and increase the prediction accuracy by 0.03–0.20, when compared with individual attributes. These results demonstrate that the predicting power can be increased significantly by combining all attributes.

For these three classifiers, decision tree has the best performance, logistic regression is the next, and naïve Bayes has the worse performance. Logistic

Table 6.1 The AUC of different classifiers built with various attributes (the bold values in the last row indicate the AUC of the classifiers built with all the above attributes)

Attributes	Logistic regression	Decision tree	Naïve Bayes
Topic attendance	0.718	0.702	0.633
Topic loyalty	0.852	0.818	0.834
Recent attendance	0.389	0.658	0.618
Weekly preference	0.752	0.746	0.752
Daily preference	0.715	0.700	0.716
Home distance	0.395	0.699	0.397
Location preference	0.671	0.685	0.670
All attributes	**0.872**	**0.875**	**0.847**

Table 6.2 The prediction accuracy of different classifiers built with various attributes (the bold values in the last row indicate the prediction accuracy of the classifiers built with all the above attributes)

Attributes	Logistic regression	Decision tree	Naïve Bayes
Topic attendance	0.639	0.656	0.602
Topic loyalty	0.760	0.763	0.736
Recent attendance	0.503	0.628	0.504
Weekly preference	0.680	0.703	0.658
Daily preference	0.656	0.660	0.654
Home distance	0.503	0.637	0.503
Location preference	0.561	0.658	0.560
All attributes	**0.786**	**0.805**	**0.767**

Table 6.3 The coefficients of attributes in the logistic regression classifier (Topic loyalty has the largest coefficient, shown in bold)

Attributes	Coefficients
Topic attendance	0.0118
Topic loyalty	**6.0789**
Recent attendance	0.0118
Weekly preference	1.95
Daily preference	1.05
Home distance	0
Location preference	0

regression and naïve Bayes underperform due to their intrinsic mechanisms in the classifiers. Logistic regression is based on a linear model, which is not sufficient to characterize the effect of each attribute. Naïve Bayes assumes that multiple attributes are independent, but the attributes may be correlated to some extent.

From the above results, we can see that combining all attributes significantly outperforms using one individual attribute. Here, we want to find out the contribution of each individual attribute when all attributes are used in the classifier. To achieve this goal, we study the inner structures of the two classifiers: logistic regression and decision tree. We will not consider naïve Bayes since it has the worst performance.

Table 6.3 shows the coefficients of the attributes (absolute value) in the logistic regression classifier. As shown in the table, *topic loyalty* has the highest contribution since it has the largest coefficient 6.0789. Therefore, when predicting whether a user will attend an event or not, the most important thing is to check whether the user is "loyal" to the event topic, i.e., whether the user has attended all or most of the events of this topic. Two other temporal attributes, *weekly preference* and *daily preference*, also have high contributions, since they have the second and the third highest coefficients, 1.95 and 1.05, respectively. This demonstrates that users' behaviors usually follow some temporal routines, especially weekly and daily patterns. From the table, we can also see that two spatial attributes *location preference* and *home distance* have 0 coefficients, which indicates that the spatial attributes do not contribute to event attendance prediction in logistic regression.

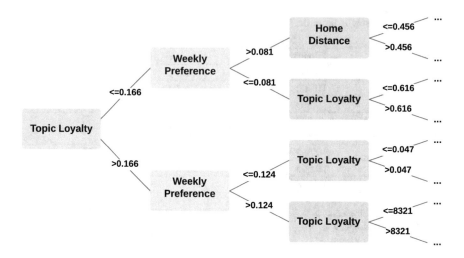

Fig. 6.9 The top three levels of decision tree

The top three levels of the decision tree are shown in Fig. 6.9. As we can see, the attribute *topic loyalty* is used in the root node and most of the nodes at the third level, demonstrating its contribution in the classifier. The attribute *weekly preference* is used in two nodes at the second level, indicating the effectiveness of using "weekly preference". We also find that the spatial attribute "home distance" is only used in one node at the third level. This verifies that spatial attributes play less important roles than the semantic and temporal attributes for event attendance prediction.

6.4.3   *The Effects of Semantic Information*

One novelty of the proposed data mining approach is to leverage the semantic information. In this subsection, we illustrate how semantic information is used to improve performance. In the first experiment, we study the impact of semantic information (i.e., semantic analysis) on individual attributes. As presented in Chap. 4, all attributes such as semantic attributes, temporal attributes and spatial attributes rely on semantic analysis. For example, the temporal attribute is calculated by averaging the temporal relations of the future event with all past events. When computing the average of the temporal relations, the event similarity between the past event and the future event is used as the weight of the past event. The spatial attribute *location preference* is also calculated using the same method. We run an experiment to see how the semantic information affects the predicting power of these attributes. Since the spatial attribute *home distance* does not use semantic information, it is not tested here. Figure 6.10 compares the predicting power of the

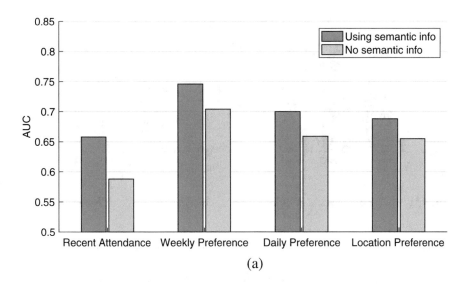

(a)

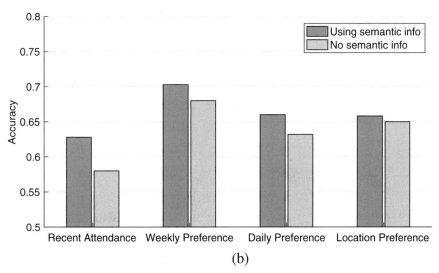

(b)

Fig. 6.10 Semantic information can be used to increase the predicting power of various attributes such as *recent attendance*, *weekly preference*, *daily preference* and *location preference* (evaluated using the decision tree). (**a**) AUC. (**b**) Accuracy

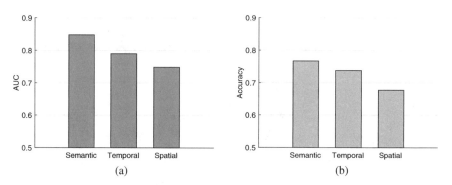

Fig. 6.11 Comparing three sets of attributes using decision tree. (**a**) AUC. (**b**) Accuracy

attributes calculated using semantic information and the attributes without semantic information. As can be seen, the attributes with semantic information can improve the AUC by up to 0.07, and improve the prediction accuracy by up to 0.05. These results demonstrate the importance of leveraging the semantic information for event attendance prediction.

In the second experiment, we compare classifiers built with three sets of attributes: semantic attributes, temporal attributes, and spatial attributes. As shown in Fig. 6.11, the classifier built with semantic attributes has the best performance in terms of AUC and prediction accuracy. This result confirms that semantic attributes are more important than temporal attributes and spatial attributes for predicting event attendance.

6.4.4   Effects of Training Set Length

In the previous experiments, the first half of the dataset is used for training and the second half is used for evaluation. We further run experiments to see if the length of the training set (as a percentage of the whole dataset) affects the performance of the supervised classifiers. The evaluation results of all three supervised classifiers are shown in Fig. 6.12. As the length of the training set increases, the result of logistic regression does not change too much, and decision tree shows some improvement. Naïve Bayes performs slightly better, but it is still the worst among the three. From the figure, we can see that decision tree has better performance when the percentage of training set is larger than 40%, while logistic regression has better performance when the percentage of training set is smaller than 40%. This result demonstrates that logistic regression only requires a small number of training data to build the classifier. If there is only limited amount of training data, logistic regression is the best classifier to choose. On the other hand, with enough training data, decision tree will be the first choice for predicting event attendance.

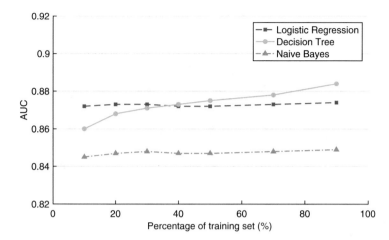

Fig. 6.12 The effect of the length of the training set on the performance of different classifiers

Chapter 7
Conclusions and Future Research Directions

Abstract This chapter concludes the book. Throughput the book, we introduced the problem of event attendance prediction, and proposed a context-aware data mining approach to solve it. Experimental results based on the collected dataset demonstrated that the proposed approach can predict event attendance with high accuracy. Finally, we point out future research directions.

In this book, we introduced the event attendance prediction problem, and proposed techniques to solve it by mining users' attendance history. Because each new event is different from past events, semantic similarity was proposed to capture the topic relevance between different events. In particular, a category-based semantic analysis method was designed to calculate the semantic similarity.

This book also presented three sets of attributes to characterize users' likelihood of attending a future event, including semantic, temporal, and spatial attributes. Semantic attributes characterize users' interest in specific topics, by studying the users' attendance to similar-topic events in the past. Semantic similarity between events was used to identify similar-topic events a user attended in the past. Temporal attributes were defined to exploit temporal information on user activities, and measure the user's temporal preference when attending events. Spatial attributes were defined to measure users' location preference when attending events.

Three supervised learning models were trained to learn how the attributes affect event attendance, including logistic regression, decision tree and naïve Bayes. To evaluate the performance of the proposed solutions, we collected a dataset based on a popularly used event-based social network, that contains semantic descriptions of all events organized over a period of 2 years. Evaluation results showed that the supervised classifiers built by all attributes outperform those built by individual attributes, and semantic attributes are more effective than temporal attributes and spatial attributes for predicting event attendance.

The research on event attendance prediction can be extended in the following directions.

- First, in addition to the supervised learning methods discussed in this book, more advanced learning methods can be used. For example, some advanced learning

© The Author(s), under exclusive license to Springer Nature Switzerland AG 2021   47
X. Zhang, G. Cao, *Event Attendance Prediction in Social Networks*, SpringerBriefs in Statistics, https://doi.org/10.1007/978-3-030-89262-3_7

methods such as support vector machines [98] and deep learning methods like neural networks [99] can be applied. Some boosting methods such as AdaBoost [100], bagging [101] can be added to the basic classifiers presented in this book to improve performance. In addition to applying different learning methods, another improvement can be extracting more attributes. For example, we can extract attributes that further consider the social influence among users. This is based on the observation that if two users usually attend the same events, it is likely they are acquaintances and may attend more events together. The new attributes can measure how the acquaintances' behaviors influence users' decisions.

Different attributes may have different effectiveness and it is important to filter out the ineffective attributes, to reduce the computation time and improve performance. Therefore, we will study attribute selection methods [102] such as correlation coefficient evaluation methods or subset selection methods, to choose the best attributes for event prediction.

- Second, COVID-19 pandemic has significantly affected people's life, especially the offline social events. During the pandemic, most social events are hosted online. The online events no longer have location information and can change users' decisions on event attendance. With online events, people are no longer confined by the geographical distance and can attend events at any location. On the other hand, people who prefer offline social atmosphere may find online events unattractive and choose not to attend. Thus, it will be interesting to see how pandemic affects users' social behaviors on event attendance. To start with, new dataset should be collected from event-based social networks during year 2020, new attributes that quantify users' online preference should be extracted, and some old attributes may have to be removed.
- Third, people's social interests usually change over time. For example, people may be interested in gym-based exercises for several weeks and then the enthusiasm fades out. A female may join mom club when she is pregnant and quit after her child grows up. Considering the change of social interests, it is important to characterize and analyze how people's event attendance behavior evolves over time, and how such evolution is related to topic. This research can benefit event and social group organizers who should adjust their plans based on the predictions.

References

1. W.-S. Soh, H.S. Kim, A predictive bandwidth reservation scheme using mobile positioning and road topology information. IEEE/ACM Trans. Netw. **14**(5), 1078–1091 (2006)
2. Q. Lv, Y. Qiao, N. Ansari, J. Liu, J. Yang, Big data driven hidden markov model based individual mobility prediction at points of interest. IEEE Trans. Vehic. Technol. **66**(6), 5204–5216 (2016)
3. X. Zhang, G. Cao, Transient community detection and its application to data forwarding in delay tolerant networks. IEEE/ACM Trans. Netw. **25**(5), 2829–2843, 2017.
4. W. Gao, G. Cao, User-centric data dissemination in disruption tolerant networks, in *IEEE International Conference on Computer Communications (Infocom)* (2011)
5. X. Zhang, G. Cao, Efficient data forwarding in mobile social networks with diverse connectivity characteristics, in *IEEE International Conference on Distributed Computing Systems (ICDCS)* (2014)
6. T.M.T. Do, D. Gatica-Perez, Contextual conditional models for smartphone-based human mobility prediction, in *ACM Conference on Ubiquitous Computing (Ubicomp)* (2012)
7. A. Noulas, S. Scellato, N. Lathia, C. Mascolo, Mining user mobility features for next place prediction in location-based services, in *IEEE International Conference on Data Mining (ICDM)* (2012)
8. B. Shen, X. Liang, Y. Ouyang, M. Liu, W. Zheng, K.M. Carley, Stepdeep: A novel spatial-temporal mobility event prediction framework based on deep neural network, in *Proceedings of the 24th ACM SIGKDD International Conference on Knowledge Discovery & Data Mining* (2018), pp. 724–733
9. W. Liu, Y. Shoji, DeepVM: RNN-based vehicle mobility prediction to support intelligent vehicle applications. IEEE Trans. Ind. Inf. **16**(6), 3997–4006 (2019)
10. C. Song, Z. Qu, N. Blumm, A.-L. Barabási, Limits of predictability in human mobility. Science **327**(5968), 1018–1021 (2010)
11. D. Yang, D. Zhang, V.W. Zheng, Z. Yu, Modeling user activity preference by leveraging user spatial temporal characteristics in LBSNs. IEEE Trans. Syst. Man Cyber. Syst. **45**(1), 129–142 (2014)
12. X. Liu, Q. He, Y. Tian, W.-C. Lee, J. McPherson, J. Han, Event-based social networks: Linking the online and offline social worlds, in *Proceedings of the 18th ACM SIGKDD International Conference on Knowledge Discovery and Data Mining* (2012), pp. 1032–1040
13. A. Mohamed, O. Onireti, S.A. Hoseinitabatabaei, M. Imran, A. Imran, R. Tafazolli, Mobility prediction for handover management in cellular networks with control/data separation, in *IEEE International Conference on Communications* (2015)

© The Author(s), under exclusive license to Springer Nature Switzerland AG 2021
X. Zhang, G. Cao, *Event Attendance Prediction in Social Networks*, SpringerBriefs in Statistics, https://doi.org/10.1007/978-3-030-89262-3

14. M. Vukovic, G. Vujnovic, D. Grubisic, Adaptive user movement prediction for advanced location-aware services, in *IEEE International Conference on Software, Telecommunications & Computer Networks* (2009)
15. W. Gao, G. Cao, Fine-grained mobility characterization: steady and transient state behaviors, in *ACM International Symposium on Mobile Ad hoc Networking and Computing (Mobihoc)* (2010)
16. S. Gambs, M.-O. Killijian, M.N. del Prado Cortez, Next place prediction using mobility markov chains, in *Proceedings of the First Workshop on Measurement, Privacy, and Mobility* (2012), pp. 1–6
17. R. Mariescu-Istodor, R. Ungureanu, P. Fränti, Real-time destination prediction for mobile users, in *15th International Conference on Location Based Services (LBS 2019)* (Copernicus GmbH, Göttingen, 2019)
18. M. Daoui, M. Belkadi, L. Chamek, M. Lalam, S. Hamrioui, A. Berqia, Mobility prediction and location management based on data mining, in *IEEE Next Generation Networks and Services* (2012)
19. Y. Jia, Y. Wang, X. Jin, X. Cheng, TSBM: The temporal-spatial bayesian model for location prediction in social networks, in *IEEE/WIC/ACM International Joint Conferences on Web Intelligence and Intelligent Agent Technologies* (2014)
20. T.V.T. Duong, D.Q. Tran, A fusion of data mining techniques for predicting movement of mobile users. J Commun. Netw. **17**(6), 568–581 (2015)
21. S.H. Ariffin, N. Abd, N.E. Ghazali, et al., Mobility prediction via markov model in LTE femtocell. Int. J. Comput. Appl. **65**(18), 40–44 (2013)
22. A. Hadachi, O. Batrashev, A. Lind, G. Singer, E. Vainikko, Cell phone subscribers mobility prediction using enhanced markov chain algorithm, in *IEEE Intelligent Vehicles Symposium Proceedings* (2014)
23. P.S. Prasad, P. Agrawal, A generic framework for mobility prediction and resource utilization in wireless networks, in *IEEE International Conference on Communication Systems and Networks* (2010)
24. S. Qiao, D. Shen, X. Wang, N. Han, W. Zhu, A self-adaptive parameter selection trajectory prediction approach via hidden markov models. IEEE Trans. Intell. Transport. Syst. **16**(1), 284–296 (2014)
25. C. Yang, M. Sun, W.X. Zhao, Z. Liu, E.Y. Chang, A neural network approach to jointly modeling social networks and mobile trajectories. ACM Trans. Inf. Syst. **35**(4), 1–28 (2017)
26. P. Wang, Y. Fu, J. Zhang, X. Li, D. Lin, Learning urban community structures: a collective embedding perspective with periodic spatial-temporal mobility graphs. ACM Trans. Intell. Syst. Technol. **9**(6), 1–28 (2018)
27. D. Zhang, K. Lee, I. Lee, Semantic periodic pattern mining from spatio-temporal trajectories. Inf. Sci. **502**, 164–189 (2019)
28. A. Sadilek, J. Krumm, Far out: Predicting long-term human mobility, in *AAAI Conference on Artificial Intelligence* (2012)
29. S. Counts, J. Geraci, Incorporating physical co-presence at events into digital social networking, in *ACM CHI Extended Abstracts on Human Factors in Computing Systems* (2005)
30. B. Xu, A. Chin, D. Cosley, On how event size and interactivity affect social networks, in *ACM CHI Extended Abstracts on Human Factors in Computing Systems* (2013)
31. S. Ricken, L. Barkhuus, Q. Jones, Going online to meet offline: Organizational practices of social activities through meetup, in *Proceedings of the 8th International Conference on Communities and Technologies* (2017), pp. 139–148
32. G. Li, Y. Liu, B. Ribeiro, H. Ding, On new group popularity prediction in event-based social networks. IEEE Trans. Netw. Sci. Eng. **7**(3), 1239–1250 (2019)
33. X. Liu, T. Suel, What makes a group fail: Modeling social group behavior in event-based social networks, in *IEEE International Conference on Big Data (Big Data)* (2016)
34. K. Feng, G. Cong, S.S. Bhowmick, S. Ma, In search of influential event organizers in online social networks, in *ACM SIGMOD International Conference on Management of Data* (2014)

35. K. Li, W. Lu, S. Bhagat, L.V. Lakshmanan, C. Yu, On social event organization, in *ACM SIGKDD International Conference on Knowledge Discovery and Data Mining* (2014)
36. J.S. Zhang, Q. Lv, Understanding event organization at scale in event-based social networks. ACM Trans. Intell. Syst. Technol. **10**(2), 1–23 (2019)
37. S. Gao, Z. Zhang, S. Su, J. Wen, L. Sun, Fair-aware competitive event influence maximization in social networks. IEEE Trans. Netw. Sci. Eng. **7**(4), 2528–2540 (2020)
38. Y. Tang, Y. Gong, L. Xu, Q. Zhang, H. Liu, S. Wang, Q. Wang, X. Gao, Is danmaku an effective way for promoting event based social network? in *ACM Conference on Computer Supported Cooperative Work and Social Computing* (2017)
39. Z. Yu, R. Du, B. Guo, H. Xu, T. Gu, Z. Wang, D. Zhang, Who should i invite for my party? Combining user preference and influence maximization for social events, in *ACM International Joint Conference on Pervasive and Ubiquitous Computing* (2015)
40. C. Ai, M. Han, J. Wang, M. Yan, An efficient social event invitation framework based on historical data of smart devices, in *IEEE International Conferences on Big Data and Cloud Computing* (2016)
41. J. She, Y. Tong, L. Chen, C.C. Cao, Conflict-aware event-participant arrangement and its variant for online setting. IEEE Trans. Knowl. Data Eng. **28**(9), 2281–2295 (2016)
42. J. She, Y. Tong, L. Chen, T. Song, Feedback-aware social event-participant arrangement, in *ACM International Conference on Management of Data* (2017)
43. Y. Cheng, Y. Yuan, L. Chen, C. Giraud-Carrier, G. Wang, Complex event-participant planning and its incremental variant, in *IEEE 33rd International Conference on Data Engineering (ICDE)* (2017)
44. H. Sun, S. Zhang, J. Huang, L. He, X. Jiang, Z. Duan, A personalized event–participant arrangement framework based on user interests in social network. Comput. Netw. **183** 107607 (2020)
45. F. Kou, Z. Zhou, H. Cheng, J. Du, Y. Shi, P. Xu, Interaction-aware arrangement for event-based social networks, in *IEEE 35th International Conference on Data Engineering (ICDE)* (2019)
46. Y. Cheng, Y. Yuan, L. Chen, C. Giraud-Carrier, G. Wang, B. Li, Event-participant and incremental planning over event-based social networks. IEEE Trans. Knowl. Data Eng. **33**, 474–488 (2019)
47. X. Liu, Y. Tian, M. Ye, W.-C. Lee, Exploring personal impact for group recommendation, in *ACM International Conference on Information and Knowledge Management* (2012)
48. Q. Yuan, G. Cong, C.-Y. Lin, COM: A generative model for group recommendation, in *ACM SIGKDD International Conference on Knowledge Discovery and Data Mining* (2014)
49. W. Zhang, J. Wang, W. Feng, Combining latent factor model with location features for event-based group recommendation, in *ACM SIGKDD International Conference on Knowledge Discovery and Data Mining* (2013)
50. S. Purushotham, C.-C. J. Kuo, Personalized group recommender systems for location-and event-based social networks. ACM Trans. Spatial Algor. Syst. **2**(4), 1–29 (2016)
51. W. Zhang, J. Wang, A collective bayesian poisson factorization model for cold-start local event recommendation, in *ACM SIGKDD International Conference on Knowledge Discovery and Data Mining* (2015)
52. Z. Wang, Y. Zhang, H. Chen, Z. Li, F. Xia, Deep user modeling for content-based event recommendation in event-based social networks, in *IEEE International Conference on Computer Communications (Infocom)* (2018)
53. M. Xu, S. Liu, Semantic-enhanced and context-aware hybrid collaborative filtering for event recommendation in event-based social networks. IEEE Access **7**, 17493–17502 (2019)
54. Y. Du, X. Meng, Y. Zhang, CVTM: A content-venue-aware topic model for group event recommendation. IEEE Trans. Knowl. Data Eng. **32**(7), 1290–1303 (2019)
55. Z. Qiao, P. Zhang, Y. Cao, C. Zhou, L. Guo, B. Fang, Combining heterogenous social and geographical information for event recommendation, in *Proceedings of the AAAI Conference on Artificial Intelligence*, vol. 28, no. 1 (2014)

56. Y. Lai, Y. Zhang, X. Meng, Y. Du, Preference and constraint factor model for event recommendation, *in IEEE Transactions on Knowledge and Data Engineering*. https://doi.org/10.1109/TKDE.2020.3046932

57. Z. Wang, Y. Zhang, Y. Li, Q. Wang, F. Xia, Exploiting social influence for context-aware event recommendation in event-based social networks, in *IEEE International Conference on Computer Communications (Infocom)* (2017)

58. P. Mahajan, P.D. Kaur, Harnessing user's social influence and IoT data for personalized event recommendation in event-based social networks. Soc. Netw. Analy. Mining **11**(1), 1–20 (2021)

59. Y. Liao, W. Lam, L. Bing, X. Shen, Joint modeling of participant influence and latent topics for recommendation in event-based social networks. ACM Trans. Inf. Syst. **36**(3), 1–31 (2018)

60. R. Mariescu-Istodor, A.S. Sayem, P. Fränti, Activity event recommendation and attendance prediction. J. Location based Serv. **13**(4), 293–319 (2019)

61. P. Fränti, K. Waga, C. Khurana, Can social network be used for location-aware recommendation?, in *WEBIST* (2015), pp. 558–565

62. P. Mahajan, P.D. Kaur, Three-tier IoT-edge-cloud (3T-IEC) architectural paradigm for real-time event recommendation in event-based social networks. J. Ambient Intell. Humaniz. Comput. **12**, 1–24 (2020)

63. T.-A.N. Pham, X. Li, G. Cong, Z. Zhang, A general graph-based model for recommendation in event-based social networks, in *IEEE 31st International Conference on Data Engineering (ICDE)* (2015)

64. S. Pramanik, R. Haldar, A. Kumar, S. Pathak, B. Mitra, Deep learning driven venue recommender for event-based social networks. IEEE Trans. Knowl. Data Eng. **32**(11), 2129–2143 (2019)

65. Y. Lu, Z. Qiao, C. Zhou, Y. Hu, L. Guo, Location-aware friend recommendation in event-based social networks: A bayesian latent factor approach, in *Proceedings of the 25th ACM International on Conference on Information and Knowledge Management* (2016), pp. 1957–1960

66. H. Yin, L. Zou, Q.V.H. Nguyen, Z. Huang, X. Zhou, Joint event-partner recommendation in event-based social networks, in *IEEE 34th International Conference on Data Engineering (ICDE)* (2018)

67. S. Karanikolaou, I. Boutsis, V. Kalogeraki, Understanding event attendance through analysis of human crowd behavior in social networks, in *Proceedings of the 8th ACM International Conference on Distributed Event-Based Systems* (2014), pp. 322–325

68. X. Zhao, T. Xu, Q. Liu, H. Guo, Exploring the choice under conflict for social event participation, in *International Conference on Database Systems for Advanced Applications* (Springer, Berlin, 2016), pp. 396–411

69. T. Trinh, D. Wu, J.Z. Huang, M. Azhar, Activeness and loyalty analysis in event-based social networks. Entropy **22**(1), 119 (2020)

70. X. Zhang, J. Zhao, G. Cao, Who will attend?–predicting event attendance in event-based social network, in *16th IEEE International Conference on Mobile Data Management* vol. 1 (2015), pp. 74–83

71. R. Du, Z. Yu, T. Mei, Z. Wang, Z. Wang, B. Guo, Predicting activity attendance in event-based social networks: Content, context and social influence, in *ACM International Joint Conference on Pervasive and Ubiquitous Computing* (2014)

72. T. Xu, H. Zhu, H. Zhong, G. Liu, H. Xiong, E. Chen, Exploiting the dynamic mutual influence for predicting social event participation. IEEE Trans. Knowl. Data Eng. **31**(6), 1122–1135 (2018)

73. X. Wu, Y. Dong, B. Shi, A. Swami, N.V. Chawla, Who will attend this event together? Event attendance prediction via deep LSTM networks, in *Proceedings of the 2018 SIAM International Conference on Data Mining* (2018), pp. 180–188

74. M. Lu, Y. Dai, H. Zhong, D. Ye, K. Yan, MFDAU: A multi-features event participation prediction model for users of different activity levels. Phys. A Statist. Mechan. Appl. **534**, 122244 (2019)
75. J. Zhang, W. Jiang, J. Zhang, J. Wu, G. Wang, Exploring weather data to predict activity attendance in event-based social network: From the organizer's view. ACM Trans. Web **15**(2), 1–25 (2021)
76. T. Chung, J. Han, D. Choi, T.T. Kwon, J.-Y. Rha, H. Kim, Privacy leakage in event-based social networks: a meetup case study. Proc. ACM Human-Comput. Int. **1** (CSCW), 1–22 (2017)
77. C. Ingram, A. Drachen, How software practitioners use informal local meetups to share software engineering knowledge, in *IEEE/ACM International Conference on Software Engineering (ICSE)* (2020), pp. 161–173
78. C.J. Date, E.F. Codd, The relational and network approaches: Comparison of the application programming interfaces, in *Proceedings of the 1974 ACM SIGFIDET (now SIGMOD) Workshop on Data Description, Access and Control: Data Models: Data-Structure-Set versus Relational* (1975), pp. 83–113
79. M. Levandowsky, D. Winter, Distance between sets. Nature **234**(5323), 34–35 (1971)
80. G.A. Miller, Wordnet: a lexical database for english. Commun. ACM **38**(11), 39–41 (1995)
81. C. Fellbaum, Wordnet, in *Theory and Applications of Ontology: Computer Applications* (Springer, Berlin, 2010), pp. 231–243
82. D. Lin, An information-theoretic definition of similarity, in *The 15th International Conference on Machine Learning (ICML)* (1998)
83. M. Rezaei, P. Fränti, Matching similarity for keyword-based clustering, in *Joint IAPR International Workshops on Statistical Techniques in Pattern Recognition (SPR) and Structural and Syntactic Pattern Recognition (SSPR)* (Springer, Berlin, 2014) pp. 193–202
84. S. Banerjee, On geodetic distance computations in spatial modeling. Biometrics **61**(2), 617–625 (2005)
85. K. Waga, A. Tabarcea, M. Chen, P. Fränti, Detecting movement type by route segmentation and classification, in *The 8th IEEE International Conference on Collaborative Computing: Networking, Applications and Worksharing (CollaborateCom)* (2012), pp. 508–513
86. I.H. Witten, E. Frank, *Data Mining: Practical Machine Learning Tools and Techniques* (Morgan Kaufmann, Burlington, 2005)
87. D.W. Hosmer Jr, S. Lemeshow, *Applied Logistic Regression* (Wiley, Hoboken, 2004)
88. J.R. Quinlan, *C4. 5: Programs for Machine Learning* (Morgan Kaufmann, Burlington, 1993)
89. I. Rish, An empirical study of the naive bayes classifier, in *IJCAI Workshop on Empirical Methods in Artificial Intelligence* (2001)
90. I.J. Myung, Tutorial on maximum likelihood estimation. J. Math. Psychol. **47**(1), 90–100 (2003)
91. J. Han, C. Moraga, The influence of the sigmoid function parameters on the speed of backpropagation learning, in *International Workshop on Artificial Neural Networks* (Springer, Berlin, 1995), pp. 195–201
92. X. Wu, V. Kumar, J.R. Quinlan, J. Ghosh, Q. Yang, H. Motoda, G.J. McLachlan, A. Ng, B. Liu, S.Y. Philip, et al., Top 10 algorithms in data mining. Knowl. Inf. Syst. **14**(1), 1–37 (2008)
93. J.R. Quinlan, Induction of decision trees. Mach. Learn. **1**(1), 81–106 (1986)
94. T. Fawcett, An introduction to roc analysis. Pattern Recog. Lett. **27**(8), 861–874 (2006)
95. A.P. Bradley, The use of the area under the roc curve in the evaluation of machine learning algorithms. Pattern Recog. **30**(7), 1145–1159 (1997)
96. J. Huang, C.X. Ling, Using auc and accuracy in evaluating learning algorithms. IEEE Trans. Knowl. Data Eng. **17**(3), 299–310 (2005)
97. W. Iba, P. Langley, Induction of one-level decision trees, in *The 9th International Conference on Machine Learning (ICML)* (1992)
98. C. Cortes, V. Vapnik, Support-vector networks. Mach. Learn. **20**(3), 273–297 (1995)

99. J. Schmidhuber, Deep learning in neural networks: an overview. Neural Netw. **61**, 85–117 (2015)

100. Y. Freund, R. Schapire, N. Abe, A short introduction to boosting. J. Jpn Soc. Artif. Intell. **14**(771–780), 1612 (1999)

101. L. Breiman, Bagging predictors. Machine Learn. **24**(2), 123–140 (1996)

102. I. Guyon, A. Elisseeff, An introduction to variable and feature selection. J. Mach. Learn. Res. **3**(Mar), 1157–1182 (2003)

Printed in the United States
by Baker & Taylor Publisher Services